Encyclopaedia of

Genes, Genomes, Genetics and Chromosomes

Encyclopaedia of Genes, Genomes, Genetics and Chromosomes

Volume 5

Editor:
Dr. R.N. Shukla

AGROTECH PRESS
JAIPUR ▪ NEW DELHI

AGROTECH PRESS

Regd. Off. "Bony Residency", Gate No. 2
Opp. Tilak Public School, Vishweswariya Nagar,
Gopalpura Road, Jaipur - 302018, India
Ph: 0141 - 2761280, Telefax: 2761381
E-mail : agrotechpress@gmail.com

Delhi Office: 102, 1st Floor, Satyam House,
4327/3, Ansari Road, Darya Ganj,
New Delhi - 110002, India
Ph: 011-45652440
visit us: www.agrotechpress.com

Distributed By
KEYA INTERNATIONAL PRESS LLP
E-mail: keyainternationalpressllp@gmail.com

Published in India, 2024

ISBN : 9789384568238 (5 Vols. Set)

Typeset by Thomson Press

Printed and bound in India by Thomson Press Ltd.,
New Delhi

Contents

Preface

A chromosome is an organized structure of DNA and protein found in cells. It is a single piece of coiled DNA containing many genes, regulatory elements and other nucleotide sequences. Chromosomes also contain DNA-bound proteins, which serve to package the DNA and control its functions. Chromosomal DNA encodes most or all of an organism's genetic information; some species also contain plasmids or other extrachromosomal genetic elements.

Chromosomes vary widely between different organisms. The DNA molecule may be circular or linear, and can be composed of 100,000 to over 3,750,000,000 nucleotides in a long chain. Typically, eukaryotic cells (cells with nuclei) have large linear chromosomes and prokaryotic cells (cells without defined nuclei) have smaller circular chromosomes, although there are many exceptions to this rule. Also, cells may contain more than one type of chromosome; for example, mitochondria in most eukaryotes and chloroplasts in plants have their own small chromosomes. In eukaryotes, nuclear chromosomes are packaged by proteins into a condensed structure called chromatin. This allows the very long DNA molecules to fit into the cell nucleus. The structure of chromosomes and chromatin varies through the cell cycle. Chromosomes are the essential unit for cellular division and must be replicated, divided, and passed successfully to their daughter cells so as to ensure the genetic diversity and survival of their progeny. Chromosomes may exist as either duplicated or unduplicated. Unduplicated chromosomes are single linear strands, whereas duplicated chromosomes contain two identical copies joined by a centromere.

Although red blood cells do not contain chromosomes, white cells do. Chromosomes can be analysed from less than one teaspoon (five millilitres) of blood, collected into a tube containing a chemical that prevents clotting. In the case of a small baby, chromosome studies can be obtained on as little as one quarter teaspoon of blood, or even on blood obtained from a fingerstick.
Amniotic fluid contains cells from the fetus, and these cells can be grown in tissue culture so that sufficient living, dividing cells are available for

chromosome analysis within 5-10 days. This is useful for diagnosis of leukemia and other bone marrow cancers, where the chromosome abnormalities are only present in the leukemia or tumour cells. Bone marrow cells can be used to provide extremely rapid chromosome diagnosis in a seriously ill infant. Sometimes a skin biopsy or tissue biopsy is used to provide material for chromosome analysis. Fetal tissue obtained after a miscarriage can also be used for chromosome analysis. Chromosome analysis is highly accurate. Not just one cell, but at least 15-20 cells are examined whenever a chromosome analysis is done. This is to determine whether *all* cells, or just some cells, have normal chromosomes. Detailed analysis at the microscope and the computer-based image analysis system are performed on at least two cells. The analysts are highly skilled cytogenetic technologists with many years of experience. Laboratory directors (The Genetics Centre has two) are licensed by New York State, and the laboratories must pass periodic inspections and proficiency testing. Very small chromosome abnormalities may not be detected by routine chromosome analysis. High resolution chromosome analysis is available, as are special staining techniques and techniques based on molecular genetic technology. Since genetic conditions may be from changes that are too small to be seen under a microscope, normal results of chromosome analysis do not guarantee that there are no genetic problems. The geneticist can discuss these issues more thoroughly, on an individual basis, after obtaining a detailed family history and reviewing any relevant medical information. Accurate interpretation is as important as an accurate result. Interpretation of abnormal chromosome results takes a great deal of skill; only a qualified clinical cytogeneticist or clinical geneticist, with a PhD or MD, should interpret the significance of chromosome results. Clinical geneticists and cytogeneticists have experience in dealing with chromosome abnormalities and can explain the clinical implications of a chromosome result.

This book provides an easy guide to difficult subject. Basic theory is linked with suggestions for suitable protocols. Techniques are explained in an easy to follow step-by-step format. It will be highly useful to the students, teachers, researchers.

—Editor

1

Understanding Genetics and Genetic Engineering

Immunology

Immunology is a broad branch of biomedical science that covers the study of all aspects of the immune system in all organisms. It deals with the physiological functioning of the immune system in states of both health and disease; malfunctions of the immune system in immunological disorders (autoimmune diseases, hypersensitivities, immune deficiency, transplant rejection); the physical, chemical and physiological characteristics of the components of the immune system in vitro, in situ, and in vivo. Immunology has applications in several disciplines of science, and as such is further divided.

Histological Examination of the Immune System

Even before the concept of immunity (from *immunis*, Latin for "exempt") was developed, numerous early physicians characterized organs that would later prove to be part of the immune system. The key primary lymphoid organs of the immune system are like thymus and bone marrow, and secondary lymphatic tissues such as spleen, tonsils, lymph vessels, lymph nodes, adenoids, and skin and liver. When health conditions warrant, immune system organs including the thymus, spleen, portions of bone marrow, lymph nodes and secondary lymphatic tissues can be surgically excised for examination while patients are still alive. Many components of the immune system are actually cellular in nature and not associated with any specific organ but rather are embedded or circulating in various tissues located throughout the body.

Classical Immunology

Classical immunology ties in with the fields of epidemiology and medicine. It studies the relationship between the body systems, pathogens, and immunity. The earliest written mention of immunity can be traced back to the plague of Athens in 430 BCE. Thucydides noted that people who had recovered from a previous bout of the disease could nurse the sick without contracting the illness a second time. Many other ancient societies have references to this phenomenon, but it was not until the 19th and 20th centuries before the concept developed into scientific theory. The study of the molecular and cellular components that comprise the immune system, including their function and interaction, is the central science of immunology. The immune system has been divided into a more primitive innate immune system, and acquired or adaptive immune system of vertebrates, the latter of which is further divided into humoral and cellular components.

The humoral (antibody) response is defined as the interaction between antibodies and antigens. Antibodies are specific proteins released from a certain class of immune cells (B lymphocytes). Antigens are defined as anything that elicits generation of antibodies, hence they are Antibody Generators. Immunology itself rests on an understanding of the properties of these two biological entities. However, equally important is the cellular response, which can not only kill infected cells in its own right, but is also crucial in controlling the antibody response. Put simply, both systems are highly interdependent. In the 21st century, immunology has broadened its horizons with much research being performed in the more specialized niches of immunology. This includes the immunological function of cells, organs and systems not normally associated with the immune system, as well as the function of the immune system outside classical models of immunity (Yemeserach 2010).

Clinical Immunology

Clinical immunology is the study of diseases caused by disorders of the immune system (failure, aberrant action, and malignant growth of the cellular elements of the system). It also involves diseases of other systems, where immune reactions play a part in the pathology and clinical features.

The diseases caused by disorders of the immune system fall into two broad categories: immunodeficiency, in which parts of the immune system fail to provide an adequate response (examples include chronic granulomatous disease), and autoimmunity, in which the immune system attacks its own host's body (examples include systemic lupus erythematosus, rheumatoid arthritis, Hashimoto's disease and myasthenia gravis). Other immune system disorders include different hypersensitivities, in which the system responds inappropriately to harmless compounds (asthma and other allergies) or responds too intensely.The study of the molecular and cellular components that comprise the immune system, including their function and interaction, is the central science of immunology. The immune system has been divided into a more primitive.

The most well-known disease that affects the immune system itself is AIDS, caused by HIV. AIDS is an immunodeficiency characterized by the lack of CD4+ ("helper") T cells and macrophages, which are destroyed by HIV. Clinical immunologists also study ways to prevent transplant rejection, in which the immune system attempts to destroy allografts os.

Developmental Immunology

The body's capability to react to antigen depends on a person's age, antigen type, maternal factors and the area where the antigen is presented. Neonates are said to be in a state of physiological immunodeficiency, because both their innate and adaptive immunological responses are greatly suppressed. Once born, a child's immune system responds favourably to protein antigens while not as well to glycoproteins and polysaccharies. In fact, many of the infections acquired by neonates are caused by low virulence organisms like Staphylococcus and Pseudomonas. In neonates, opsonic activity and the ability to activate the complement cascade is very limited.

For example, the mean level of C3 in a newborn is approximately 65% of that found in the adult. Phagocytic activity is also greatly impaired in newborns. This is due to lower opsonic activity, as well as diminished up-regulation of integrin and selectin receptors, which limit the ability of neutrophils to interact with adhesion molecules in the endothelium. Their monocytes are slow and have a reduced ATP production, which also limits the newborns phagocitic activity. Although, the number of total lymphocytes is significantly higher than in adults, the cellular and humoral immunity is also impaired. Antigen presenting cells in newborns have a reduced capability to activate T cells. Also, T cells of a newborn proliferate poorly and produce very small amounts of cytokines like IL-2, IL-4, IL-5, IL-12, and IFN-g which limits their capacity to activate the humoral response as well as the phagocitic activity of macrophage. B cells develop early in gestation but are not fully active. Maternal factors also play a role in the body's immune response. At birth most of the immunoglobulin is present is maternal IgG. Because IgM, IgD, IgE and IgA don't cross the placenta, they are almost undetectable at birth. Although some IgA is provided in breast milk. These passively acquired antibodies can protect the newborn up to 18 months, but their response is usually short-lived and of low affinity. These antibodies can also produce a negative response. If a child is exposed to the antibody for a particular antigen before being exposed to the antigen itself then the child will produce a dampened response.

Passively acquired maternal antibodies can suppress the antibody response to active immunization. Similarly the response of T-cells to vaccination differs in children compared to adults, and vaccines that induce Th1 responses in adults do not readily elicit these same responses in neonates. By 6-9 months after birth, a child's immune system begins to respond more strongly to glycoproteins. Not until 12-24 months of age is there a marked improvement in the body's response to polysaccharides. This can be the reason for the specific time frames found in vaccination schedules.

During adolescence the human body undergoes several physical, physiological and immunological changes. These changes are started and mediated by different hormones. Depending on the sex either testosterone or 17-â-oestradiol, act on male and female bodies accordingly, start acting at ages of 12 and 10 years. There is evidence that these steroids act directly not only on the primary and secondary sexual characteristics, but also have an effect on the development and regulation of the immune system. There is an increased risk in developing autoimmunity for pubescent and post pubescent females and males. There is also some evidence that cell surface receptors on B cells and macrophages may detect sex hormones in the system.

The female sex hormone 17-â-oestradiol has been shown to regulate the level of immunological response. Similarly, some male androgens, like testosterone, seem to suppress the stress response to infection; but other androgens like DHEA have the opposite effect, as it increases the immune response instead of down playing it. As in females, the male sex hormones seem to have more control of the immune system during puberty and the time right after than in fully developed adults. Other than hormonal changes physical changes like the involution of the Thymus during puberty will also affect the immunological response of the subject or patient.

Immunotherapy

The use of immune system components to treat a disease or disorder is known as immunotherapy. Immunotherapy is most commonly used in the context of the treatment of cancers together with chemotherapy (drugs) and radiotherapy (radiation). However, immunotherapy is also often used in the immunosuppressed (such as HIV patients) and people suffering from other immune deficiencies or autoimmune diseases.

Diagnostic Immunology

The specificity of the bond between antibody and antigen has made it an excellent tool in the detection of substances in a variety of diagnostic techniques. Antibodies specific for a desired antigen can be conjugated with a radiolabel, fluorescent label, or colour-forming enzyme and are used as a "probe" to detect it. However, the similarity between some antigens can lead to false positives and other errors in such tests by antibodies cross-reacting with antigens that aren't exact matches.

Evolutionary Immunology

Study of the immune system in extant species is capable of giving us a key understanding of the evolution of species and the immune system. A development of complexity of the immune system can be seen from simple phagocytotic protection of single celled organisms, to circulating antimicrobial peptides in insects to lymphoid organs in vertebrates. However, it is important to recognize that every organism living today has an immune system that has evolved to be absolutely capable of protecting it from most forms of harm; those organisms that did not adapt their immune systems to external threats are no longer around to be observed. Insects and other arthropods, while not possessing true adaptive immunity, show highly evolved systems of innate immunity, and are additionally protected from external injury (and exposure to pathogens) by their chitinous shells.

Reproductive Immunology

This area of the immunology is devoted to the study of immunological aspects of the reproductive process including fetus acceptance. The term has also been used by fertility clinics to address fertility problems, recurrent miscarriages, premature deliveries, and dangerous complications such as pre-eclampsia.

Immunologist

According to the American Academy of Allergy, Asthma, and Immunology (AAAAI), "an immunologist is a research scientist who investigates the immune system of vertebrates (including the human immune system). Immunologists include research scientists (Ph.D.) who work in laboratories. Immunologists also include physicians who, for example, treat patients with immune system disorders. Some immunologists are physician-scientists who combine laboratory research with patient care."

Genetics

Genetics, a discipline of biology, is the science of genes, heredity, and variation in living organisms. Genetics deals with the molecular structure and function of genes, with gene behaviour in the context of a cell or organism (e.g. dominance and epigenetics), with patterns of inheritance from parent to offspring, and with gene distribution, variation and change in populations. Given that genes are universal to living organisms, genetics can be applied to the study of all living systems, from viruses and bacteria, through plants (especially crops), to humans, as in medical genetics.

The fact that living things inherit traits from their parents has been used since prehistoric times to improve crop plants and animals through selective breeding. However, the modern science of genetics, which seeks to understand the process of inheritance, only began with the work of Gregor Mendel in the mid-19th century. Although he did not know the physical basis for heredity, Mendel observed that organisms inherit traits via discrete units of inheritance, which are now called genes. Genes correspond to regions within DNA, a molecule composed of a chain of four different types of nucleotides—the sequence of these nucleotides is the genetic information organisms inherit. DNA naturally occurs in a double stranded form, with nucleotides on each strand complementary to each other. Each strand can act as a template for creating a new partner strand. This is the physical method for making copies of genes that can be inherited.

The sequence of nucleotides in a gene is translated by cells to produce a chain of amino acids, creating proteins—the order of amino acids in a protein corresponds to the order of nucleotides in the gene. This relationship between nucleotide sequence and amino acid sequence is known as the genetic code. The amino acids in a protein determine how it folds into a three-dimensional shape; this structure is, in turn, responsible for the protein's function. Proteins carry out almost all the functions needed for cells to live. A change to the DNA in a gene can change a protein's amino acds, changing its shape and function: this can have a dramatic effect in the cell and on the organism as a whole. Although genetics plays a large role in the appearance and behaviour of organisms, it is the combination of genetics with what an organism experiences that determines the ultimate outcome. For example, while genes play a role in determining an organism's size, the nutrition and health it experiences after inception also have a large effect.

History

Although the science of genetics began with the applied and theoretical work of Gregor Mendel in the mid-19th century, other theories of inheritance preceded Mendel. A popular theory during Mendel's time was the concept of blending inheritance: the idea that individuals inherit a smooth blend of traits from their parents. Mendel's work disproved this, showing that traits are composed of combinations of distinct genes rather than a continuous blend. Another theory that had some support at that time was the inheritance of acquired characteristics: the belief that individuals inherit traits strengthened by their parents. This theory (commonly associated with Jean-Baptiste Lamarck) is now known to be wrong—the experiences of individuals do not affect the genes they pass to their children. Other theories included the pangenesis of Charles Darwin (which had both acquired and inherited aspects) and Francis Galton's reformulation of pangenesis as both particulate and inherited.

Mendelian and Classical Genetics: Modern genetics started with Gregor Johann Mendel, a German-Czech Augustinian monk and scientist who studied the nature of inheritance in plants. In his paper "Versuche über Pflanzenhybriden" ("Experiments on Plant Hybridization"), presented in 1865 to the *Naturforschender Verein* (Society for Research in Nature) in Brünn, Mendel traced the inheritance patterns of certain traits in pea plants and described them mathematically. Although this pattern of inheritance could only be observed for a few traits, Mendel's work suggested that heredity was particulate, not acquired, and that the inheritance patterns of many traits could be explained through simple rules and ratios. The importance of Mendel's work did not gain wide understanding until the 1890s, after his death, when other scientists working on similar problems re-discovered his research. William Bateson, a proponent of Mendel's work, coined the word *genetics* in 1905. (The adjective *genetic*, derived from the Greek word *genesis*—ãÝíåóéò, "origin", predates the noun and was first used in a biological sense in 1860.) Bateson popularized the usage of the word *genetics* to describe the study of inheritance in his inaugural address to the Third International Conference on Plant Hybridization in London, England, in 1906.

After the rediscovery of Mendel's work, scientists tried to determine which molecules in the cell were responsible for inheritance. In 1910, Thomas Hunt Morgan argued that genes are on chromosomes, based on observations of a sex-linked white eye mutation in fruit flies. In 1913, his student Alfred Sturtevant used the phenomenon of genetic linkage to show that genes are arranged linearly on the chromosome.

Molecular Genetics

Although genes were known to exist on chromosomes, chromosomes are composed of both protein and DNA—scientists did not know which of these is responsible for inheritance. In 1928, Frederick Griffith discovered the phenomenon of transformation: dead bacteria could transfer genetic material to "transform" other still-living bacteria. Sixteen years later, in 1944, Oswald Theodore Avery, Colin McLeod and Maclyn McCarty identified the molecule responsible for transformation as DNA. The Hershey-Chase experiment in 1952 also showed that DNA (rather than protein) is the genetic material of the viruses that infect bacteria, providing further evidence that DNA is the molecule responsible for inheritance.

James D. Watson and Francis Crick determined the structure of DNA in 1953, using the X-ray crystallography work of Rosalind Franklin and Maurice Wilkins that indicated DNA had a helical structure (i.e., shaped like a corkscrew). Their double-helix model had two strands of DNA with the nucleotides pointing inward, each matching a complementary nucleotide on the other strand to form what looks like rungs on a twisted ladder. This structure showed that genetic information exists in the sequence of nucleotides on each strand of DNA. The structure also suggested a simple method for duplication: if the strands are separated, new partner strands can be reconstructed for each based on the sequence of the old strand.

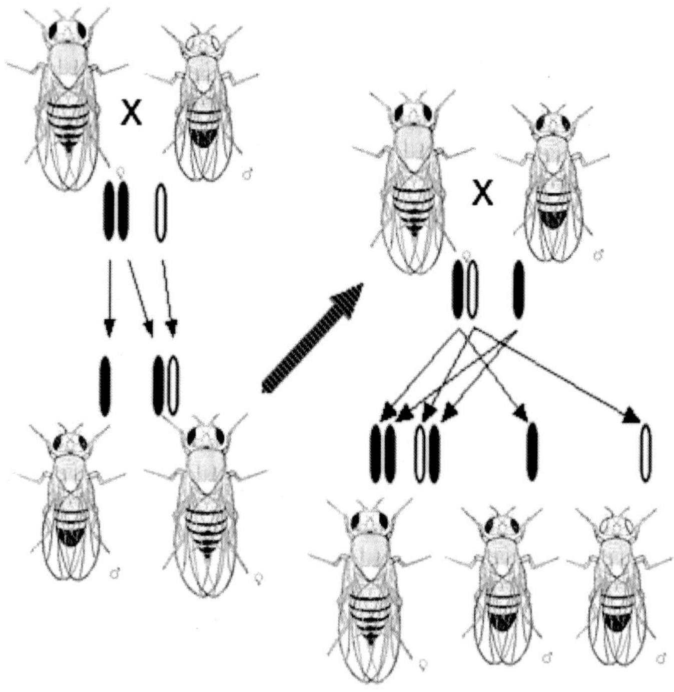

Figure: *Morgan's observation of sex-linked inheritance of a mutation causing white eyes in Drosophila led him to the hypothesis that genes are located upon chromosomes.*

Although the structure of DNA showed how inheritance works, it was still not known how DNA influences the behaviour of cells. In the following years, scientists tried to understand how DNA controls the process of protein production. It was discovered that the cell uses DNA as a template to create matching messenger RNA (a molecule with nucleotides, very similar to DNA). The nucleotide sequence of a messenger RNA is used to create an amino acid sequence in protein; this translation between nucleotide and amino acid sequences is known as the genetic code.

With this molecular understanding of inheritance, an explosion of research became possible. One important development was chain-termination DNA sequencing in 1977 by Frederick Sanger. This technology allows scientists to read the nucleotide sequence of a DNA molecule. In 1983, Kary Banks Mullis developed the polymerase chain reaction, providing a quick way to isolate and amplify a specific section of a DNA from a mixture. Through the pooled efforts of the Human Genome Project and the parallel private effort by Celera Genomics, these and other techniques culminated in the sequencing of the human genome in 2003.

Features of Inheritance

Discrete Inheritance and Mendel's Laws: At its most fundamental level, inheritance in organisms occurs by means of discrete traits, called genes. This property was first observed by Gregor Mendel, who studied the segregation of heritable traits in pea plants. In his experiments studying the trait for flower colour, Mendel observed that the flowers of each pea plant were either purple or white—but never an intermediate between the two colours. These different, discrete versions of the same gene are called alleles.

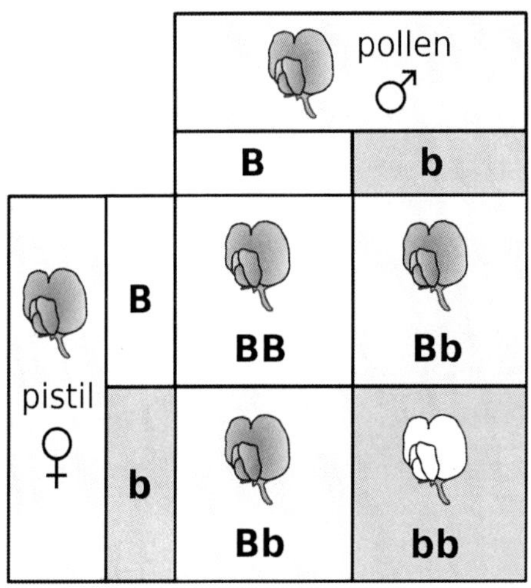

Figure: *A Punnett square depicting a cross between two pea plants heterozygous for purple (B) and white (b) blossoms*

In the case of pea, which is a diploid species, each individual plant has two alleles of each gene, one allele inherited from each parent. Many species, including humans, have this pattern of inheritance. Diploid organisms with two copies of the same allele of a given gene are called homozygous at that gene locus, while organisms with two different alleles of a given gene are called heterozygous.

The set of alleles for a given organism is called its genotype, while the observable traits of the organism are called its phenotype. When organisms are heterozygous at a gene, often one allele is called dominant as its qualities dominate the phenotype of the organism, while the other allele is called recessive as its qualities recede and are not observed. Some alleles do not have complete dominance and instead have incomplete dominance by expressing an intermediate phenotype, or codominance by expressing both alleles at once.

When a pair of organisms reproduce sexually, their offspring randomly inherit one of the two alleles from each parent. These observations of discrete inheritance and the segregation of alleles are collectively known as Mendel's first law or the Law of Segregation.

Notation and Diagrams

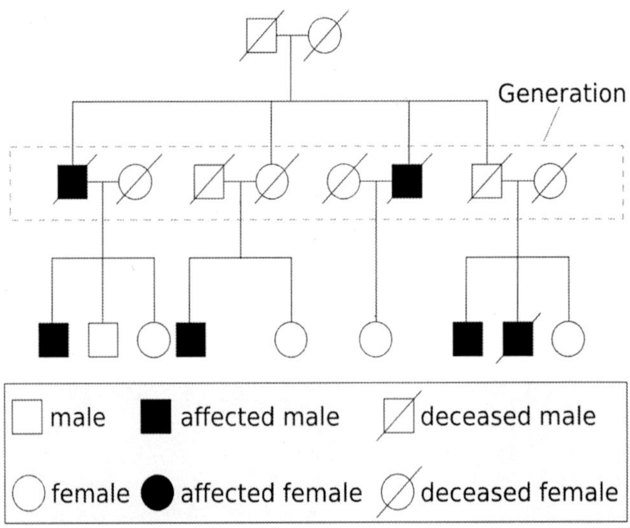

Figure: Genetic pedigree charts help track the inheritance patterns of traits.

Geneticists use diagrams and symbols to describe inheritance. A gene is represented by one or a few letters. Often a "+" symbol is used to mark the usual, non-mutant allele for a gene. In fertilization and breeding experiments (and especially when discussing Mendel's laws) the parents are referred to as the "P" generation and the offspring as the "F1" (first filial) generation. When the F1 offspring mate with each other, the offspring are called the "F2" (second filial) generation. One of the common diagrams used to predict the result of cross-breeding is the Punnett square. When studying human genetic diseases, geneticists often use pedigree charts to represent the inheritance of traits. These charts map the inheritance of a trait in a family tree.

Interactions of Multiple Genes

Organisms have thousands of genes, and in sexually reproducing organisms these genes generally assort independently of each other. This means that the inheritance of an allele for yellow or green pea colour is unrelated to the inheritance of alleles for white or purple flowers. This phenomenon, known as "Mendel's second law" or the "Law of independent assortment", means that the alleles of different genes get shuffled between parents to form offspring with many different combinations. (Some genes do not assort independently, demonstrating genetic linkage, a topic discussed later in this article.)

Often different genes can interact in a way that influences the same trait. In the Blue-eyed Mary (*Omphalodes verna*), for example, there exists a gene with alleles that determine the colour of flowers: blue or magenta. Another gene, however, controls whether the flowers have colour at all or are white. When a plant has two copies of this white allele, its flowers are white—regardless of whether the first gene has blue or magenta alleles. This interaction between genes is called epistasis, with the second gene epistatic to the first.

Many traits are not discrete features (e.g. purple or white flowers) but are instead continuous features (e.g. human height and skin colour). These complex traits are products of many genes. The influence of these genes is mediated, to varying degrees, by the environment an organism has experienced. The degree to which an organism's genes contribute to a complex trait is called heritability. Measurement of the heritability of a trait is relative—in a more variable environment, the environment has a bigger influence on the total variation of the trait. For example, human height is a trait with complex causes. It has a heritability of 89% in the United States. In Nigeria, however, where people experience a more variable access to good nutrition and health care, height has a heritability of only 62%.

Molecular Basis for Inheritance

DNA and Chromosomes: The molecular basis for genes is deoxyribonucleic acid (DNA). DNA is composed of a chain of nucleotides, of which there are four types: adenine (A), cytosine (C), guanine (G), and thymine (T). Genetic information exists in the sequence of these nucleotides, and genes exist as stretches of sequence along the DNA chain. Viruses are the only exception to this rule—sometimes viruses use the very similar molecule RNA instead of DNA as their genetic material. DNA normally exists as a double-stranded molecule, coiled into the shape of a double-helix. Each nucleotide in DNA preferentially pairs with its partner nucleotide on the opposite strand: A pairs with T, and C pairs with G. Thus, in its two-stranded form, each strand effectively contains all necessary information, redundant with its partner strand. This structure of DNA is the physical basis for inheritance: DNA replication duplicates the genetic information by splitting the strands and using each strand as a template for synthesis of a new partner strand.

Genes are arranged linearly along long chains of DNA sequence, called chromosomes. In bacteria, each cell usually contains a single circular chromosome, while eukaryotic organisms (including plants and animals) have their DNA arranged in multiple linear chromosomes. These DNA strands are often extremely long; the largest human chromosome, for example, is about 247 million base pairs in length. The DNA of a chromosome is associated with structural proteins that organize, compact, and control access to the DNA, forming a material called chromatin; in eukaryotes, chromatin is usually composed of nucleosomes, segments of DNA wound around cores of histone proteins. The full set of hereditary material in an organism (usually the combined DNA sequences of all chromosomes) is called the genome.

While haploid organisms have only one copy of each chromosome, most animals and many plants are diploid, containing two of each chromosome and thus two copies of every gene. The two alleles for a gene are located on identical loci of sister chromatids, each allele inherited from a different parent. Many species have so called sex chromosomes. They are special in that they determine the sex of the organism. In humans and many other animals, the Y-chromosome contains the gene that triggers the development of the specifically male characteristics. In evolution, this chromosome has lost most of its content and also most of its genes, while the X chromosome is similar to the other chromosomes and contains many genes. The X and Y chromosomes form a very heterogeneous pair before cell division.

Reproduction: When cells divide, their full genome is copied and each daughter cell inherits one copy. This process, called mitosis, is the simplest form of reproduction and is the basis for asexual reproduction. Asexual reproduction can also occur in multicellular organisms, producing offspring that inherit their genome from a single parent. Offspring that are genetically identical to their parents are called clones. Eukaryotic organisms often use sexual reproduction to generate offspring that contain a mixture of genetic material inherited from two different parents. The process of sexual reproduction alternates between forms that contain single copies of the genome (haploid) and double copies (diploid). Haploid cells fuse and combine genetic material to create a diploid cell with paired chromosomes. Diploid organisms form haploids by dividing, without replicating their DNA, to create daughter cells that randomly inherit one of each pair of chromosomes. Most animals and many plants are diploid for most of their lifespan, with the haploid form reduced to single cell gametes such as sperm or eggs. Although they do not use the haploid/diploid method of sexual reproduction, bacteria have many methods of acquiring new genetic information. Some bacteria can undergo conjugation, transferring a small circular piece of DNA to another bacterium. Bacteria can also take up raw DNA fragments found in the environment and integrate them into their genomes, a phenomenon known as transformation. These processes result in horizontal gene transfer, transmitting fragments of genetic information between organisms that would be otherwise unrelated.

Recombination and Linkage

The diploid nature of chromosomes allows for genes on different chromosomes to assort independently during sexual reproduction, recombining to form new combinations of genes. Genes on the same chromosome would theoretically never recombine, however, were it not for the process of chromosomal crossover. During crossover, chromosomes exchange stretches of DNA, effectively shuffling the gene alleles between the chromosomes. This process of chromosomal crossover generally occurs during meiosis, a series of cell divisions that creates haploid cells.

The probability of chromosomal crossover occurring between two given points on the chromosome is related to the distance between the points. For an arbitrarily long distance, the probability of crossover is high enough that the inheritance of the genes is effectively uncorrelated. For genes that are closer together, however, the lower probability of crossover means that the genes demonstrate genetic linkage—alleles for the two genes tend to be inherited together. The amounts of linkage between a series of genes can be combined to form a linear linkage map that roughly describes the arrangement of the genes along the chromosome.

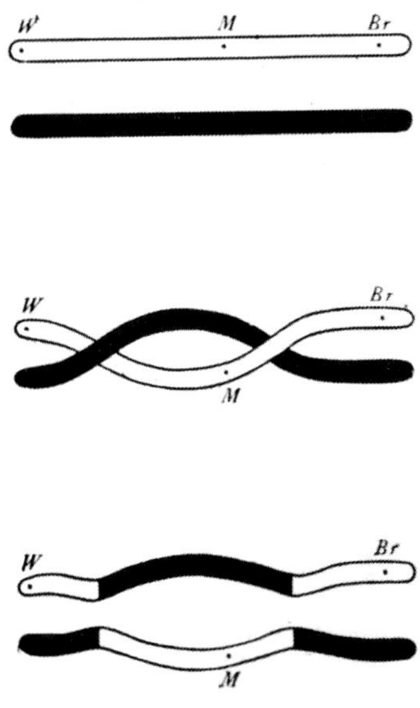

Figure: *Thomas Hunt Morgan's 1916 illustration of a double crossover between chromosomes*

Gene Expression

Genetic Code: Genes generally express their functional effect through the production of proteins, which are complex molecules responsible for most functions in the cell. Proteins are chains of amino acids, and the DNA sequence of a gene (through an RNA intermediate) is used to produce a specific protein sequence. This process begins with the production of an RNA molecule with a sequence matching the gene's DNA sequence, a process called transcription.

This messenger RNA molecule is then used to produce a corresponding amino acid sequence through a process called translation. Each group of three nucleotides in the sequence, called a codon, corresponds either to one of the twenty possible amino acids in a protein or an instruction to end the amino acic sequence; this correspondence is called the genetic code. The flow of information is unidirectional: information is transferred from nucleotide sequences into the amino acid sequence of proteins, but it never transfers from protein back into the sequence of DNA—a phenomenon Francis Crick called the central dogma of molecular biology. The specific sequence of amino acids results in a unique three-dimensional structure for that protein, and the three-dimensional structures of proteins are related to their functions. Some are simple structural molecules, like the fibres formed by the protein collagen.

Proteins can bind to other proteins and simple molecules, sometimes acting as enzymes by facilitating chemical reactions within the bound molecules (without changing the structure of the protein itself). Protein structure is dynamic; the protein hemoglobin bends into slightly different forms as it facilitates the capture, transport, and release of oxygen molecules within mammalian blood. A single nucleotide difference within DNA can cause a change in the amino acid sequence of a protein. Because protein structures are the result of their amino acid sequences, some changes can dramatically change the properties of a protein by destabilizing the structure or changing the surface of the protein in a way that changes its interaction with other proteins and molecules. For example, sickle-cell anemia is a human genetic disease that results from a single base difference within the coding region for the â-globin section of hemoglobin, causing a single amino acid change that changes hemoglobin's physical properties. Sickle-cell versions of hemoglobin stick to themselves, stacking to form fibres that distort the shape of red blood cells carrying the protein. These sickle-shaped cells no longer flow smoothly through blood vessels, having a tendency to clog or degrade, causing the medical problems associated with this disease. Some genes are transcribed into RNA but are not translated into protein products—such RNA molecules are called non-coding RNA. In some cases, these products fold into structures which are involved in critical cell functions (e.g. ribosomal RNA and transfer RNA). RNA can also have regulatory effect through hybridization interactions with other RNA molecules (e.g. microRNA).

Nature Versus Nurture: Although genes contain all the information an organism uses to function, the environment plays an important role in determining the ultimate phenotype—a phenomenon often referred to as "nature vs. nurture". The phenotype of an organism depends on the interaction of genetics with the environment. One example of this is the case of temperature-sensitive mutations. Often, a single amino acid change within the sequence of a protein does not change its behaviour and interactions with other molecules, but it does destabilize the structure. In a high temperature environment, where molecules are moving more quickly and hitting each other, this results in the protein losing its structure and failing to function. In a low temperature environment, however, the protein's structure is stable and it functions normally. This type of mutation is visible in the coat colouration of Siamese cats, where a mutation in an enzyme responsible for pigment production causes it to destabilize and lose function at high temperatures. The protein remains functional in areas of skin that are colder—legs, ears, tail, and face—and so the cat has dark fur at its extremities.

Environment also plays a dramatic role in effects of the human genetic disease phenylketonuria. The mutation that causes phenylketonuria disrupts the ability of the body to break down the amino acid phenylalanine, causing a toxic build-up of an intermediate molecule that, in turn, causes severe symptoms of progressive mental retardation and seizures. If someone with the phenylketonuria mutation follows a strict diet that avoids this amino acid, however, they remain normal and healthy. A popular method to determine how much role nature and nurture play is to study identical and fraternal twins or siblings of multiple birth. Because identical siblings come from the same zygote they are genetically the same. Fraternal siblings however are as different genetically from one another as normal siblings. By comparing how often the twin of a set has the same disorder between fraternal and identical twins, scientists can see whether there is more of a nature or nurture effect. One famous example of a multiple birth study includes the Genain quadruplets, who were identical quadruplets all diagnosed with schizophrenia.

Gene Regulation: The genome of a given organism contains thousands of genes, but not all these genes need to be active at any given moment. A gene is expressed when it is being transcribed into mRNA (and translated into protein), and there exist many cellular methods of controlling the expression of genes such that proteins are produced only when needed by the cell. Transcription factors are regulatory proteins that bind to the start of genes, either promoting or inhibiting the transcription of the gene. Within the genome of *Escherichia coli* bacteria, for example, there exists a series of genes necessary for the synthesis of the amino acid tryptophan. However, when tryptophan is already available to the cell, these genes for tryptophan synthesis are no longer needed. The presence of tryptophan directly affects the activity of the genes—tryptophan molecules bind to the tryptophan repressor (a transcription factor), changing the repressor's structure such that the repressor binds to the genes. The tryptophan repressor blocks the transcription and expression of the genes, thereby creating negative feedback regulation of the tryptophan synthesis process.

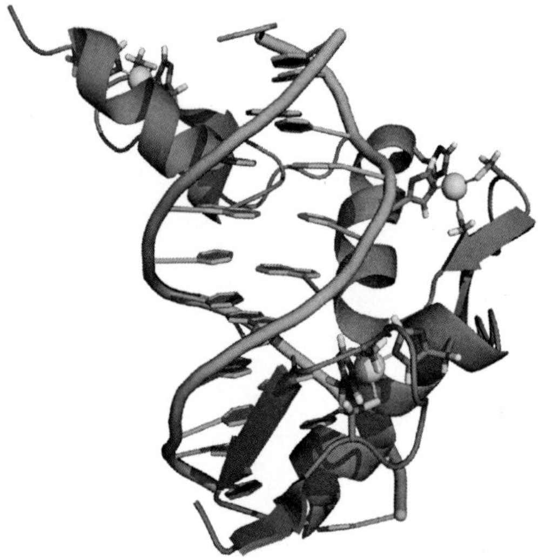

Figure: *Transcription factors bind to DNA, influencing the transcription of associated genes.*

Differences in gene expression are especially clear within multicellular organisms, where cells all contain the same genome but have very different structures and behaviours due to the expression of different sets of genes. All the cells in a multicellular organism derive from a single cell, differentiating into variant cell types in response to external and intercellular signals and gradually establishing different patterns of gene expression to create different behaviours. As no single gene is responsible for the development of structures within multicellular organisms, these patterns arise from the complex interactions between many cells.

Within eukaryotes there exist structural features of chromatin that influence the transcription of genes, often in the form of modifications to DNA and chromatin that are stably inherited by daughter cells. These features are called "epigenetic" because they exist "on top" of the DNA sequence and retain inheritance from one cell generation to the next. Because of epigenetic features, different cell types grown within the same medium can retain very different properties. Although epigenetic features are generally dynamic over the course of development, some, like the phenomenon of paramutation, have multigenerational inheritance and exist as rare exceptions to the general rule of DNA as the basis for inheritance.

Genetic Change

Mutations: During the process of DNA replication, errors occasionally occur in the polymerization of the second strand. These errors, called mutations, can have an impact on the phenotype of an organism, especially if they occur within the protein coding sequence of a gene. Error rates are usually very low—1 error in every 10–100 million bases—due to the "proofreading" ability of DNA polymerases. (Without proofreading error rates are a thousandfold higher; because many viruses rely on DNA and RNA polymerases that lack proofreading ability, they experience higher mutation rates.)

Processes that increase the rate of changes in DNA are called mutagenic: mutagenic chemicals promote errors in DNA replication, often by interfering with the structure of base-pairing, while UV radiation induces mutations by causing damage to the DNA structure. Chemical damage to DNA occurs naturally as well, and cells use DNA repair mechanisms to repair mismatches and breaks in DNA—nevertheless, the repair sometimes fails to return the DNA to its original sequence.

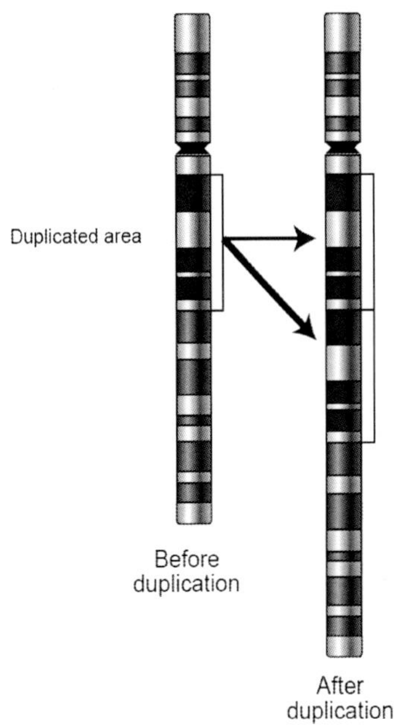

Figure: *Gene duplication allows diversification by providing redundancy: one gene can mutate and lose its original function without harming the organism.*

In organisms that use chromosomal crossover to exchange DNA and recombine genes, errors in alignment during meiosis can also cause mutations. Errors in crossover are especially likely when similar sequences cause partner chromosomes to adopt a mistaken alignment; this makes some regions in genomes more prone to mutating in this way.

These errors create large structural changes in DNA sequence—duplications, inversions or deletions of entire regions, or the accidental exchanging of whole parts between different chromosomes (called translocation).

Natural Selection and Evolution: Mutations alter an organisms genotype and occasionally this causes different phenotypes to appear. Most mutations have little effect on an organism's phenotype, health, or reproductive fitness. Mutations that do have an effect are usually deleterious, but occasionally some can be beneficial. Studies in the fly *Drosophila melanogaster* suggest that if a mutation changes a protein produced by a gene, about 70 percent of these mutations will be harmful with the remainder being either neutral or weakly beneficial.

Population genetics studies the distribution of genetic differences within populations and how these distributions change over time. Changes in the frequency of an allele in a population are mainly influenced by natural selection, where a given allele provides a selective or reproductive advantage to the organism, as well as other factors such as genetic drift, artificial selection and migration. Over many generations, the genomes of organisms can change significantly, resulting in the phenomenon of evolution. Selection for beneficial mutations can cause a species to evolve into forms better able to survive in their environment, a process called adaptation. New species are formed through the process of speciation, often caused by geographical separations that prevent populations from exchanging genes with each other. The application of genetic principles to the study of population biology and evolution is referred to as the modern synthesis.

By comparing the homology between different species' genomes it is possible to calculate the evolutionary distance between them and when they may have diverged (called a molecular clock). Genetic comparisons are generally considered a more accurate method of characterizing the relatedness between species than the comparison of phenotypic characteristics. The evolutionary distances between species can be used to form evolutionary trees; these trees represent the common descent and divergence of species over time, although they do not show the transfer of genetic material between unrelated species (known as horizontal gene transfer and most common in bacteria).

Research and Technology

Model Organisms: Although geneticists originally studied inheritance in a wide range of organisms, researchers began to specialize in studying the genetics of a particular subset of organisms. The fact that significant research already existed for a given organism would encourage new researchers to choose it for further study, and so eventually a few model organisms became the basis for most genetics research. Common research topics in model organism genetics include the study of gene regulation and the involvement of genes in development and cancer. Organisms were chosen, in part, for convenience—short generation times and easy genetic manipulation made some organisms popular genetics research tools. Widely used model organisms include the gut bacterium *Escherichia coli*, the plant *Arabidopsis thaliana*, baker's yeast (*Saccharomyces cerevisiae*), the nematode *Caenorhabditis elegans*, the common fruit fly (*Drosophila melanogaster*), and the common house mouse (*Mus musculus*).

Medicine: Medical genetics seeks to understand how genetic variation relates to human health and disease. When searching for an unknown gene that may be involved in a disease, researchers commonly use genetic linkage and genetic pedigree charts to find the location on the genome associated with the disease. At the population level, researchers take advantage of Mendelian randomization to look for locations in the genome that are associated with diseases, a technique especially useful for multigenic traits not clearly defined by a single gene. Once a candidate gene is found, further research is often done on the corresponding gene (called an orthologous gene) in model organisms. In addition to studying genetic diseases, the increased availability of genotyping techniques has led to the field of pharmacogenetics—studying how genotype can affect drug responses. Individuals differ in their inherited tendency to develop cancer, and cancer is a genetic disease.

The process of cancer development in the body is a combination of events. Mutations occasionally occur within cells in the body as they divide. Although these mutations will not be inherited by any offspring, they can affect the behaviour of cells, sometimes causing them to grow and divide more frequently. There are biological mechanisms that attempt to stop this process; signals are given to inappropriately dividing cells that should trigger cell death, but sometimes additional mutations occur that cause cells to ignore these messages. An internal process of natural selection occurs within the body and eventually mutations accumulate within cells to promote their own growth, creating a cancerous tumor that grows and invades various tissues of the body.

Research Techniques: DNA can be manipulated in the laboratory. Restriction enzymes are commonly used enzymes that cut DNA at specific sequences, producing predictable fragments of DNA. DNA fragments can be visualized through use of gel electrophoresis, which separates fragments according to their length. The use of ligation enzymes allows DNA fragments to be connected, and by ligating fragments of DNA together from different sources, researchers can create recombinant DNA.

Often associated with genetically modified organisms, recombinant DNA is commonly used in the context of plasmids—short circular DNA fragments with a few genes on them. By inserting plasmids into bacteria and growing those bacteria on plates of agar (to isolate clones of bacteria cells), researchers can clonally amplify the inserted fragment of DNA (a process known as molecular cloning). (Cloning can also refer to the creation of clonal organisms, through various techniques.) DNA can also be amplified using a procedure called the polymerase chain reaction (PCR). By using specific short sequences of DNA, PCR can isolate and exponentially amplify a targeted region of DNA. Because it can amplify from extremely small amounts of DNA, PCR is also often used to detect the presence of specific DNA sequences.

DNA Sequencing and Genomics

One of the most fundamental technologies developed to study genetics, DNA sequencing allows researchers to determine the sequence of nucleotides in DNA fragments. Developed in 1977 by Frederick Sanger and coworkers, chain-termination sequencing is now routinely used to sequence DNA fragments. With this technology, researchers have been able to study the molecular sequences associated with many human diseases. As sequencing has become less expensive, researchers have sequenced the genomes of many organisms, using computational tools to stitch together the sequences of many different fragments (a process called genome assembly). These technologies were used to sequence the human genome, leading to the completion of the Human Genome Project in 2003. New high-throughput sequencing technologies are dramatically lowering the cost of DNA sequencing, with many researchers hoping to bring the cost of resequencing a human genome down to a thousand dollars.

The large amount of sequence data available has created the field of genomics, research that uses computational tools to search for and analyze patterns in the full genomes of organisms. Genomics can also be considered a subfield of bioinformatics, which uses computational approaches to analyze large sets of biological data.

Genetic Engineering

Genetic engineering, also called genetic modification, is the direct human manipulation of an organism's genetic material in a way that does not occur under natural conditions. It involves the use of recombinant DNA techniques, but does not include traditional animal and plant breeding or mutagenesis. Any organism that is generated using these techniques is considered to be a genetically modified organism. The first organisms genetically engineered were bacteria in 1973 and then mice in 1974. Insulin producing bacteria were commercialized in 1982 and genetically modified food has been sold since 1994. The most common form of genetic engineering involves the insertion of new genetic material at an unspecified location in the host genome. This is accomplished by isolating and copying the genetic material of interest, generating a construct containing all the genetic elements for correct expression, and then inserting this construct into the host organism. Other forms of genetic engineering include gene targeting and knocking out specific genes via engineered nucleases such as zinc finger nucleases or engineered homing endonucleases.

Genetic engineering techniques have been applied in numerous fields including research, biotechnology, and medicine. Medicines such as insulin and human growth hormone are now produced in bacteria, experimental mice such as the oncomouse and the knockout mouse are being used for research purposes and insect resistant and/or herbicide tolerant crops have been commercialized. Genetically engineered plants and animals capable of producing biotechnology drugs more cheaply than current methods (called pharming) are also being developed and in 2009 the FDA approved the sale of the pharmaceutical protein antithrombin produced in the milk of genetically engineered goats.

Definition

Genetic engineering alters the genetic makeup of an organism using techniques that introduce heritable material prepared outside the organism either directly into the host or into a cell that is then fused or hybridized with the host. This involves using recombinant nucleic acid (DNA or RNA) techniques to form new combinations of heritable genetic material followed by the incorporation of that material either indirectly through a vector system or directly through micro-injection, macro-injection and micro-encapsulation techniques. Genetic engineering does not include traditional animal and plant breeding, in vitro fertilisation, induction of polyploidy, mutagenesis and cell fusion techniques that do not use recombinant nucleic acids or a genetically modified organism in the process. Cloning and stem cell research, although not considered genetic engineering, are closely related and genetic engineering can be used within them. Synthetic biology is an emerging discipline that takes genetic engineering a step further by introducing artificially synthesized genetic material from raw materials into an organism. If genetic material from another species is added to the host, the resulting organism is called transgenic. If genetic material from the same species or a species that can naturally breed with the host is used the resulting organism is called cisgenic. Genetic engineering can also be used to remove genetic material from the target organism, creating a knock out organism. In Europe genetic modification is synonymous with genetic engineering while within the United States of America it can also refer to conventional breeding methods.

History

Humans have altered the genomes of species for thousands of years through artificial selection and more recently mutagenesis. Genetic engineering as the direct manipulation of DNA by humans outside breeding and mutations has only existed since the 1970s. The term "genetic engineering" was first coined by Jack Williamson in his science fiction novel *Dragon's Island*, published in 1951, one year before DNA's role in heredity was confirmed by Alfred Hershey and Martha Chase, and two years before James Watson and Francis Crick showed that the DNA molecule has a double-helix structure.

In 1972 Paul Berg created the first recombinant DNA molecules by combined DNA from the monkey virus SV40 with that of the lambda virus. In 1973 Herbert Boyer and Stanley Cohen created the first transgenic organism by inserting antibiotic resistance genes into the plasmid of an *E. coli* bacterium. A year later Rudolf Jaenisch created a transgenic mouse by introducing foreign DNA into its embryo, making it the world's first transgenic animal. In 1976 Genentech, the first genetic engineering company was founded by Herbert Boyer and Robert Swanson and a year later and the company produced a human protein (somatostatin) in *E.coli*. Genentech announced the production of genetically engineered human insulin in 1978. In 1980, the U.S. Supreme Court in the Diamond v. Chakrabarty case ruled that genetically altered life could be patented. The insulin produced by bacteria, branded humulin, was approved for release by the Food and Drug Administration in 1982.

The first field trials of genetically engineered plants occurred in France and the USA in 1986, tobacco plants were engineered to be resistant to herbicides. The People's Republic of China was the first country to commercialize transgenic plants, introducing a virus-resistant tobacco in 1992. In 1994 Calgene attained approval to commercially release the Flavr Savr tomato, a tomato engineered to have a longer shelf life. In 1994, the European Union approved tobacco engineered to be resistant to the herbicide bromoxynil, making it the first genetically engineered crop commercialized in Europe. In 1995, Bt Potato was approved safe by the Environmental Protection Agency, making it the first pesticide producing crop to be approved in the USA. In 2009 11 transgenic crops were grown commercially in 25 countries, the largest of which by area grown were the USA, Brazil, Argentina, India, Canada, China, Paraguay and South Africa. In 2010, scientists at the J. Craig Venter Institute, announced that they had created the first synthetic bacterial genome, and added it to a cell containing no DNA. The resulting bacterium, named Synthia, was the world's first synthetic life form.

Process

Isolating the Gene: First, the gene to be inserted into the genetically modified organism must be chosen and isolated. Presently, most genes transferred into plants provide protection against insects or tolerance to herbicides. In animals the majority of genes used are growth hormone genes. Once chosen the genes must be isolated. This typically involves multiplying the gene using polymerase chain reaction (PCR). If the chosen gene or the donor organism's genome has been well studied it may be present in a genetic library. If the DNA sequence is known, but no copies of the gene are available, it can be artificially synthesized. Once isolated, the gene is inserted into a bacterial plasmid.

Constructs: The gene to be inserted into the genetically modified organism must be combined with other genetic elements in order for it to work properly. The gene can also be modified at this stage for better expression or effectiveness. As well as the gene to be inserted most constructs contain a promoter and terminator region as well as a selectable marker gene. The promoter region initiates transcription of the gene and can be used to control the location and level of gene expression, while the terminator region ends transcription. The selectable marker, which in most cases confers antibiotic resistance to the organism it is expressed in, is needed to determine which cells are transformed with the new gene. The constructs are made using recombinant DNA techniques, such as restriction digests, ligations and molecular cloning.

Gene Targeting: The most common form of genetic engineering involves inserting new genetic material randomly within the host genome. Other techniques allow new genetic material to be inserted at a specific location in the host genome or generate mutations at desired genomic loci capable of knocking out endogenous genes. The technique of gene targeting uses homologous recombination to target desired changes to a specific endogenous gene. This tends to occur at a relatively low frequency in plants and animals and generally requires the use of selectable markers. The frequency of gene targeting can be greatly enhanced with the use of engineered nucleases such as zinc finger nucleases, engineered homing endonucleases, or nucleases created from TAL effectors. In addition to enhancing gene targeting, engineered nucleases can also be used to introduce mutations at endogenous genes that generate a gene knockout.

Transformation: About 1% of bacteria are naturally able to take up foreign DNA but it can also be induced in other bacteria. Stressing the bacteria for example, with a heat shock or an electric shock, can make the cell membrane permeable to DNA that may then incorporate into their genome or exist as extrachromosomal DNA. DNA is generally inserted into animal cells using microinjection, where it can be injected through the cells nuclear envelope directly into the nucleus or through the use of viral vectors. In plants the DNA is generally inserted using *Agrobacterium*-mediated recombination or biolistics.

In *Agrobacterium*-mediated recombination the plasmid construct must also contain T-DNA. *Agrobacterium* naturally inserts DNA from a tumor inducing plasmid into any susceptible plant's genome it infects, causing crown gall disease. The T-DNA region of this plasmid is responsible for insertion of the DNA. The genes to be inserted are cloned into a binary vector, which contains T-DNA and can be grown in both *E. Coli* and *Agrobacterium*. Once the binary vector is constructed the plasmid is transformed into *Agrobacterium* containing no plasmids and plant cells are infected. The *Agrobacterium* will then naturally insert the genetic material into the plant cells. In biolistics particles of gold or tungsten are coated with DNA and then shot into young plant cells or plant embryos. Some genetic material will enter the cells and transform them. This method can be used on plants that are not susceptible to *Agrobacterium* infection and also allows transformation of plant plastids. Another transformation method for plant and animal cells is electroporation. Electroporation involves subjecting the plant or animal cell to an electric shock, which can make the cell membrane permeable to plasmid DNA. In some cases the electroporated cells will incorporate the DNA into their genome. Due to the damage caused to the cells and DNA the transformation efficiency of biolistics and electroporation is lower than agrobacterial mediated transformation and microinjection.

Selection: Not all the organism's cells will be transformed with the new genetic material; in most cases a selectable marker is used to differentiate transformed from untransformed cells. If a cell has been successfully transformed with the DNA it will also contain the marker gene. By growing the cells in the presence of an antibiotic or chemical that selects or marks the cells expressing that gene it is possible to separate the transgenic events from the non-transgenic. Another method of screening involves using a DNA probe that will only stick to the inserted gene. A number of strategies have been developed that can remove the selectable marker from the mature transgenic plant.

Regeneration: As often only a single cell is transformed with genetic material the organism must be regrown from that single cell. As bacteria consist of a single cell and reproduce clonally regeneration is not necessary. In plants this is accomplished through the use of tissue culture. Each plant species has different requirements for successful regeneration through tissue culture. If successful an adult plant is produced that contains the transgene in every cell. In animals it is necessary to ensure that the inserted DNA is present in the embryonic stem cells. When the offspring is produced they can be screened for the presence of the gene. All offspring from the first generation will be heterozygous for the inserted gene and must be mated together to produce a homozygous animal.

Confirmation: Further tests using PCR, Southern Blots and Bioassays are needed to confirm that the gene is expressed and functions correctly. The organism's offspring are also tested to ensure that the trait can be inherited and that it follows a Mendelian inheritance pattern.

Applications

Genetic engineering has applications in medicine, research, industry and agriculture and can be used on a wide range of plants, animals and micro organism.

Medicine: In medicine genetic engineering has been used to mass-produce insulin, human growth hormones, follistim (for treating infertility), human albumin, monoclonal antibodies, antihemophilic factors, vaccines and many other drugs. Vaccination generally involves injecting weak live, killed or inactivated forms of viruses or their toxins into the person being immunized. Genetically engineered viruses are being developed that can still confer immunity, but lack the infectious sequences. Mouse hybridomas, cells fused together to create monoclonal antibodies, have been humanised through genetic engineering to create human monoclonal antibodies. Genetic engineering is used to create animal models of human diseases. Genetically modified mice are the most common genetically engineered animal model. They have been used to study and model cancer (the oncomouse), obesity, heart disease, diabetes, arthritis, substance abus, anxiety, aging and Parkinson disease. Potential cures can be tested against these mouse models. Also genetically modified pigs have been bred with the aim of increasing the success of pig to human organ transplantation.

Gene therapy is the genetic engineering of humans by replacing defective human genes with functional copies. This can occur in somatic tissue or germline tissue. If the gene is inserted into the germline tissue it can be passed down to that person's descendants. Gene therapy has been used to treat patients suffering from immune deficiencies (notably Severe combined immunodeficiency) and trials have been carried out on other genetic disorders. The success of gene therapy so far has been limited and a patient (Jesse Gelsinger) has died during a clinical trial testing a new treatment. There are also ethical concerns should the technology be used not just for treatment, but for enhancement, modification or alteration of a human beings' appearance, adaptability, intelligence, character or behaviour. The distinction between cure and enhancement can also be difficult to establish. Transhumanists consider the enhancement of humans desirable.

Research: Genetic engineering is an important tool for natural scientists. Genes and other genetic information from a wide range of organisms are transformed into bacteria for storage and modification, creating genetically modified bacteria in the process. Bacteria are cheap, easy to grow, clonal, multiply quickly, relatively easy to transform and can be stored at -80°C almost indefinitely. Once a gene is isolated it can be stored inside the bacteria providing an unlimited supply for research. Organisms are genetically engineered to discover the functions of certain genes. This could be the effect on the phenotype of the organism, where the gene is expressed or what other genes it interacts with. These experiments generally involve loss of function, gain of function, tracking and expression.

- Loss of function experiments, such as in a gene knockout experiment, in which an organism is engineered to lack the activity of one or more genes. A knockout experiment involves the creation and manipulation of a DNA construct *in vitro*, which, in a simple knockout, consists of a copy of the desired gene, which has been altered such that it is non-functional. Embryonic stem cells incorporate the altered gene, which replaces the already present functional copy. These stem cells are injected into blastocysts, which are implanted into surrogate mothers. This allows the experimenter to analyze the defects caused by this mutation and thereby determine the role of particular genes. It is used especially frequently in developmental biology. Another method, useful in organisms such as Drosophila (fruit fly), is to induce mutations in a large population and then screen the progeny for the desired mutation. A similar process can be used in both plants and prokaryotes.
- Gain of function experiments, the logical counterpart of knockouts. These are sometimes performed in conjunction with knockout experiments to more finely establish the function of the desired gene. The process is much the same as that in knockout engineering, except that the construct is designed to increase the function of the gene, usually by providing extra copies of the gene or inducing synthesis of the protein more frequently.
- Tracking experiments, which seek to gain information about the localization and interaction of the desired protein. One way to do this is to replace the wild-type gene with a 'fusion' gene, which is a juxtaposition of the wild-type gene with a reporting element such as green fluorescent protein (GFP) that will allow easy visualization of the products of the genetic modification. While this is a useful technique, the manipulation can destroy the function of the gene, creating secondary effects and possibly calling into question the results of the experiment. More sophisticated techniques are now in development that can track protein products without mitigating their function, such as the addition of small sequences that will serve as binding motifs to monoclonal antibodies.
- Expression studies aim to discover where and when specific proteins are produced. In these experiments, the DNA sequence before the DNA that codes for a protein, known as a gene's promoter, is reintroduced into an organism with the protein coding region replaced by a reporter gene such as GFP or an enzyme that catalyzes the production of a dye.

Thus the time and place where a particular protein is produced can be observed. Expression studies can be taken a step further by altering the promoter to find which pieces are crucial for the proper expression of the gene and are actually bound by transcription factor proteins; this process is known as promoter bashing.

Industrial: By engineering genes into bacterial plasmids it is possible to create a biological factory that can produce proteins and enzymes. Some genes do not work well in bacteria, so yeast, a eukaryote, can also be used. Bacteria and yeast factories have been used to produce medicines such as insulin, human growth hormone, and vaccines, supplements such as tryptophan, aid in the production of food (chymosin in cheese making) and fuels. Other applications involving genetically engineered bacteria being investigated involve making the bacteria perform tasks outside their natural cycle, such as cleaning up oil spills, carbon and other toxic waste.

Agriculture: One of the best-known and controversial applications of genetic engineering is the creation of genetically modified food. There are three generations of genetically modified crops. First generation crops have been commercialized and most provide protection from insects and/or resistance to herbicides. There are also fungal and virus resistant crops developed or in development. They have been developed to make the insect and weed management of crops easier and can indirectly increase crop yield. The second generation of genetically modified crops being developed aim to directly improve yield by improving salt, cold or drought tolerance and to increase the nutritional value of the crops.

The third generation consists of pharmaceutical crops, crops that contain edible vaccines and other drugs. Some agriculturally important animals have been genetically modified with growth hormones to increase their size while others have been engineered to express drugs and other proteins in their milk. The genetic engineering of agricultural crops can increase the growth rates and resistance to different diseases caused by pathogens and parasites. This is beneficial as it can greatly increase the production of food sources with the usage of fewer resources that would be required to host the world's growing populations.

These modified crops would also reduce the usage of chemicals, such as fertilizers and pesticides, and therefore decrease the severity and frequency of the damages produced by these chemical pollution. Ethical and safety concerns have been raised around the use of genetically modified food. A major safety concern relates to the human health implications of eating genetically modified food, in particular whether toxic or allergic reactions could occur. Gene flow into related non-transgenic crops, off target effects on beneficial organisms and the impact on biodiversity are important environmental issues.

Ethical concerns involve religious issues, corporate control of the food supply, intellectual property rights and the level of labeling needed on genetically modified products.

Other Uses: In materials science, a genetically modified virus has been used to construct a more environmentally friendly lithium-ion battery.

Some bacteria have been genetically engineered to create black and white photographs while others have potential to be used as sensors by expressing a fluorescent protein under certain environmental conditions. Genetic engineering is also being used to create BioArt and novelty items such as blue roses, and glowing fish.

Opposition and Criticism

A 2010 study of Canola found transgenes in 80% of wild (uncultivated or "feral") varieties in North Dakota, meaning 80% of the plants which had established themselves in the area were genetically engineered varieties.

The researchers stated that "we found the highest densities of [such transgene-containing] plants near agricultural fields and along major freeways, but we were also finding plants in the middle of nowhere" adding that "over time,..the build-up of different types of herbicide resistance in feral [natural] canola and closely related weeds, like field mustard, could make it more difficult to manage these plants using herbicides."

Ethanol Fermentation

Ethanol fermentation, also referred to as alcoholic fermentation, is a biological process in which sugars such as glucose, fructose, and sucrose are converted into cellular energy and thereby produce ethanol and carbon dioxide as metabolic waste products.

Because yeasts perform this conversion in the absence of oxygen, ethanol fermentation is classified as anaerobic.

Figure: *Grapes fermenting during the production of wine.*

Ethanol fermentation occurs in the production of alcoholic beverages and ethanol fuel, and in the rising of bread dough.

The Chemical Process of Fermentation of Glucose

The chemical equation below summarizes the fermentation of glucose, whose chemical formula is $C_6H_{12}O_6$. One mole of glucose is converted into two moles of ethanol and two moles of carbon dioxide:

$$C_{12}H_{22}O_{11} + H_2O + \text{invertase} \rightarrow C_6H_{12}O_6$$

$$C_6H_{12}O_6 + \text{Zymase} \rightarrow 2C_2H_5OH + 2CO_2$$

C_2H_5OH is the chemical formula for ethanol.

Before fermentation takes place, one glucose molecule is broken down into two pyruvate molecules. This is known as glycolysis. Glycolysis is summarized by the chemical equation:

$$C_6H_{12}O_6 + 2\ ADP + 2\ P_i + 2\ NAD^+ \rightarrow 2\ CH_3COCOO^-$$
$$+ 2\ ATP + 2\ NADH + 2\ H_2O + 2H^+$$

The chemical formula of pyruvate is CH_3COCOO^-. P_i stands for the inorganic phosphate. As shown by the reaction equation, glycolysis causes the reduction of two molecules of NAD^+ to NADH. Two ADP molecules are also converted to two ATP and two water molecules via substrate-level phosphorylation.

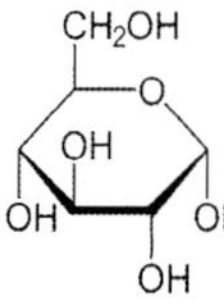

Glucose depicted in Haworth projection

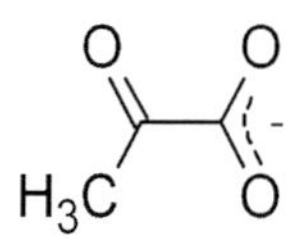

Pyruvate

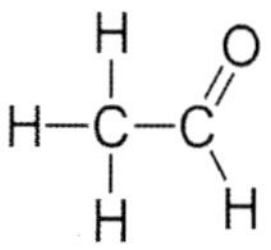

Acetaldehyde

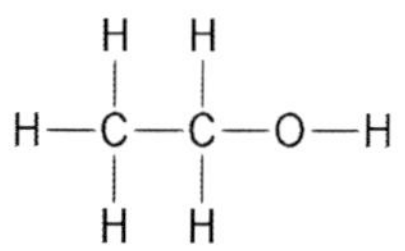

Ethanol

Effect of Oxygen

The fermentation process does not require oxygen. If oxygen is present, some species of yeast (*Kluyveromyces lactis*, *Kluyveromyces lipolytica*) oxidize pyruvate completely to carbon dioxide and water. This process is called respiration. Thus these yeasts produce ethanol only in an anaerobic environment. However, many yeasts such as the commonly used baker's yeast *Saccharomyces cerevisiae*, and *Schizosaccharomyces pombe*, prefer fermentation to respiration. These yeasts will produce ethanol even under aerobic conditions given the right sources of nutrition.

Uses

Ethanol fermentation is responsible for the rising of bread dough. Yeast organisms consume sugars in the dough and produce ethanol and carbon dioxide as waste products. The carbon dioxide forms bubbles in the dough, expanding it into something of a foam. Nearly all the ethanol evaporates from the dough when the bread is baked. All alcoholic beverages, including those produced by carbonic maceration, are produced by ethanol fermentation by yeast. Wine and brandy are produced by fermentation of the natural sugars present in fruits, especially grapes.

Beer and whiskey are produced by fermentation of grain starches that have been converted to sugar by the enzyme amylase, which is present in grain kernels that have been germinated. Amylase-treated grain or amylase-treated potatoes are fermented for the production of vodka. Rum is produced by fermentation of cane sugar. In all cases, the fermentation must take place in a vessel that allows carbon dioxide to escape, but prevents outside air from coming in, as exposure to oxygen would prevent the formation of ethanol.

Similar yeast fermentation of various carbohydrate products is used to produce much of the ethanol used for fuel.

Feedstocks for Fuel Production

The dominant ethanol feedstock in warmer regions is sugarcane. In temperate regions, sugar beet is sometimes used instead. In the United States, the main feedstock for the production of ethanol is currently corn. Approximately 2.8 gallons of ethanol are produced from one bushel of corn (0.42 liter per kilogram). While much of the corn turns into ethanol, some of the corn also yields by-products such as DDGS (distillers dried grains with solubles) that can be used as feed for livestock. A bushel of corn produces about 18 pounds of DDGS (320 kilograms of DDGS per metric ton of maize).

Although most of the fermentation plants have been built in corn-producing regions, sorghum is also an important feedstock for ethanol production in the Plains states. Pearl millet is showing promise as an ethanol feedstock for the southeastern U.S. and the potential of duckweed is being studied.

In some parts of Europe, particularly France and Italy, grapes have become a *de facto* feedstock for fuel ethanol by the distillation of surplus wine. In Japan, it has been proposed to use rice normally made into sake as an ethanol source.

Cassava as Ethanol Feedstock: Ethanol can be made from mineral oil or from sugars or starches. Starches are cheapest. The starchy crop with highest energy content per acre is cassava, which grows in tropical countries. Thailand already had a large cassava industry in the 1990s, for use as cattle feed and as a cheap admixture to wheat flour. Nigeria and Ghana are already establishing cassava-to-ethanol plants.

Production of ethanol from cassava is currently economically feasible when crude oil prices are above US$120 per barrel. New varieties of cassava are being developed, so the future situation remains uncertain. Currently, cassava can yield between 25-40 tonnes per hectare (with irrigation and fertilizer), and from a tonne of cassava roots, circa 200 liters of ethanol can be produced (assuming cassava with 22% starch content). A liter of ethanol contains circa 21.46 MJ of energy.

The overall energy efficiency of cassava-root to ethanol conversion is circa 32%. The yeast used for processing cassava is *Endomycopsis fibuligera*, sometimes used together with bacterium *Zymomonas mobilis.*

Byproducts of Fermentation

Ethanol fermentation produces unharvested byproducts such as heat, food for livestock, and water.

Microbes Used in Ethanol Fermentation

- Yeast
- Zymomonas mobilis

2

Genetics, Heredity and Variations

Genetic Linkage

Genetic linkage is a term which describes the tendency of certain loci or alleles to be inherited together. Genetic loci on the same chromosome are physically close to one another and tend to stay together during meiosis, and are thus genetically *linked*.

Background

At the beginning of normal meiosis, a chromosome pair (made up of a chromosome from the mother and a chromosome from the father) intertwine and exchange sections or fragments of chromosome. The pair then breaks apart to form two chromosomes with a new combination of genes that differs from the combination supplied by the parents.

Through this process of recombining genes, organisms can produce offspring with new combinations of maternal and paternal traits that may contribute to or enhance survival. This crossing over of DNA can cause alleles previously on the same chromosome to be separated and end up in different daughter cells. The further the two alleles are apart, the greater the chance that a cross-over event may occur between them, possibly separating the alleles.

The relative distance between two genes can be calculated using the offspring of an organism showing two linked genetic traits, and finding the percentage of the offspring where the two traits do not run together. The higher the percentage of descendants that does not show both traits, the farther apart on the chromosome the two genes are.

Among individuals of an experimental population or species, some phenotypes or traits occur randomly with respect to one another in a manner known as independent assortment. Today scientists understand that independent assortment occurs when the genes affecting the phenotypes are found on different chromosomes or separated by a great enough distance on the same chromosome that recombination occurs at least half of the time. An exception to independent assortment develops when genes appear near one another on the same chromosome. When genes occur on the same chromosome, they are usually inherited as a single unit. Genes inherited in this way are said to be linked, and are referred to as "linkage groups." For example, in fruit flies the genes affecting eye colour and wing length are inherited together because they appear on the same chromosome. Genetic linkage was first discovered by the British geneticists William Bateson and Reginald Punnett shortly after Mendel's laws were rediscovered. The understanding of genetic linkage was expanded by the work of Thomas Hunt Morgan. Morgan's observation that the amount of crossing over between linked genes differs led to the idea that crossover frequency might indicate the distance separating genes on the chromosome. Alfred Sturtevant, a student of Morgan's, first developed genetic maps, also known as linkage maps. Sturtevant proposed that the greater the distance between linked genes, the greater the chance that non-sister chromatids would cross over in the region between the genes. By working out the number of recombinants it is possible to obtain a measure for the distance between the genes. This distance is called a genetic map unit (m.u.), or a centimorgan and is defined as the distance between genes for which one product of meiosis in 100 is recombinant. A recombinant frequency (RF) of 1 % is equivalent to 1 m.u. But this equivalence is only a good approximate for small percentages; the largest percentage of recombinants cannot exceed 50%, which would be the situation where the two genes are at the extreme opposite ends of the same chromosomes.

In this situation, any crossover events would result in an exchange of genes, but only an odd number of crossover events (a 50-50 chance between even and odd number of crossover events) would result in a recombinant product of meiotic crossover. A statistical interpretation of this is through the Haldane mapping function or the Kosambi mapping function, among others. A linkage map is created by finding the map distances between a number of traits that are present on the same chromosome, ideally avoiding having significant gaps between traits to avoid the inaccuracies that will occur due to the possibility of multiple recombination events.

Linkage Map

A linkage map is a genetic map of a species or experimental population that shows the position of its known genes or genetic markers relative to each other in terms of recombination frequency, rather than as specific physical distance along each chromosome. Linkage mapping is critical for identifying the location of genes that cause genetic diseases. A genetic map is a map based on the frequencies of recombination between markers during crossover of homologous chromosomes. The greater the frequency of recombination (segregation) between two genetic markers, the farther apart they are assumed to be. Conversely, the lower the frequency of recombination between the markers, the smaller the physical distance between them. Historically, the markers originally used were detectable phenotypes (enzyme production, eye colour) derived from coding DNA sequences; eventually, confirmed or assumed noncoding DNA sequences such as microsatellites or those generating restriction fragment length polymorphisms (RFLPs) have been used.

Genetic maps help researchers to locate other markers, such as other genes by testing for genetic linkage of the already known markers. A genetic map is not a physical map (such as a radiation reduced hybrid map) or gene map. The LOD score (logarithm (base 10) of odds) is a statistical test often used for linkage analysis in human populations, and also in animal and plant populations. The LOD score compares the likelihood of obtaining the test data if the two loci are indeed linked, to the likelihood of observing the same data purely by chance. Positive LOD score favour the presence of linkage, whereas negative LOD scores indicate that linkage is less likely. The test was developed by Newton E. Morton. Computerized LOD score analysis is a simple way to analyze complex family pedigrees in order to determine the linkage between Mendelian traits (or between a trait and a marker, or two markers).

The method is described in greater detail by Strachan and Read. Briefly, it works as follows:

1. Establish a pedigree
2. Make a number of estimates of recombination frequency
3. Calculate a LOD score for each estimate
4. The estimate with the highest LOD score will be considered the best estimate.

The LOD score is calculated as follows:

NR denotes the number of non-recombinant offspring, and R denotes the number of recombinant offspring. The reason 0.5 is used in the denominator is that any alleles that are completely unlinked (e.g. alleles on separate chromosomes) have a 50% chance of recombination, due to independent assortment.

Theta is the recombinant fraction, it is equal to R/(NR + R). In practice, LOD scores are looked up in a table which lists LOD scores for various standard pedigrees and various values of recombination frequency. By convention, a LOD score greater than 3.0 is considered evidence for linkage. A LOD score of +3 indicates 1000 to 1 odds that the linkage being observed did not occur by chance. On the other hand, a LOD score less than-2.0 is considered evidence to exclude linkage. Although it is very unlikely that a LOD score of 3 would be obtained from a single pedigree, the mathematical properties of the test allow data from a number of pedigrees to be combined by summing the LOD scores. It is important to keep in mind that this traditional cutoff of LOD>+3 is an arbitrary one and that the difference between certain types of linkage studies, particularly analyses of complex genetic traits with hundreds of markers, these criteria should probably be modified to a somewhat higher cutoff.

Recombination frequency is the frequency that a single chromosomal crossover will take place between two genes during meiosis. Double crossovers would turn into no recombination. In this case we cannot tell if crossovers took place. If the loci we're analysing are very close (less than 7 cM) a double crossover is very unlikely. When distances become higher, the likelihood of a double crossover increases. Recombination frequency is a measure of genetic linkage and is used in the creation of a genetic linkage map. A centimorgan (cM) is a unit that describes a recombination frequency of 1%. In this way we can measure the genetic distance between two loci, based upon their recombination frequency. This is a good estimate of the real distance. As the likelihood of a double crossover increases we systematically underestimate the genetic distance between two loci.

During meiosis, chromosomes assort randomly into gametes, such that the segregation of alleles of one gene is independent of alleles of another gene. This is stated in Mendel's Second Law and is known as the law of independent assortment. The law of independent assortment always holds true for genes that are located on different chromosomes, but for genes that are on the same chromosome, it does not always hold true. As an example of independent assortment, consider the crossing of the pure-bred homozygote parental strain with genotype *AABB* with a different pure-bred strain with genotype *aabb*. A and a and B and b represent the alleles of genes A and B. Crossing these homozygous parental strains will result in F1 generation offspring with genotype AaBb.

The F1 offspring AaBb produces gametes that are *AB*, *Ab*, *aB*, and *ab* with equal frequencies (25%) because the alleles of gene A assort independently of the alleles for gene B during meiosis. Note that 2 of the 4 gametes (50 %)—*Ab* and *aB*—were not present in the parental generation. These gametes represent recombinant gametes. Recombinant gametes are those gametes that differ from both of the haploid gametes that made up the diploid cell. In this example, the recombination frequency is 50% since 2 of the 4 gametes were recombinant gametes. The recombination frequency will be 50% when two genes are located on different chromosomes or when they are widely separated on the same chromosome. This is a consequence of independent assortment. When two genes are close together on the same chromosome, they do not assort independently and are said to be linked. Whereas genes located on different chromosomes assort independently and have a recombination frequency of 50%, linked genes have a recombination frequency that is less than 50%.

As an example of linkage, consider the classic experiment by William Bateson and Reginald Punnett. They were interested in trait inheritance in the sweet pea and were studying two genes—the gene for flower colour (*P*, purple, and *p*, red) and the gene affecting the shape of pollen grains (*L*, long, and *l*, round). They crossed the pure lines *PPLL* and *ppll* and then self-crossed the resulting *PpLl* lines. According to Mendelian genetics, the expected phenotypes would occur in a 9:3:3:1 ratio of PL:Pl:pL:pl. To their surprise, they observed an increased frequency of PL and pl and a decreased frequency of Pl and pL. Their experiment revealed linkage between the *P* and *L* alleles and the *p* and *l* alleles. The frequency of *P* occurring together with *L* and with *p* occurring together with *l* is greater than that of the recombinant *Pl* and *pL*. The recombination frequency cannot be computed directly from this experiment, but intuitively it is less than 50%.

The progeny in this case received two dominant alleles linked on one chromosome (referred to as coupling or cis arrangement). However, after crossover, some progeny could have received one parental chromosome with a dominant allele for one trait (eg Purple) linked to a recessive allele for a second trait (eg round) with the opposite being true for the other parental chromosome (eg red and Long). This is referred to as repulsion or a trans arrangement. The phenotype here would still be purple and long but a test cross of this individual with the recessive parent would produce progeny with much greater proportion of the two crossover phenotypes. While such a problem may not seem likely from this example, unfavourable repulsion linkages do appear when breeding for disease resistance in some crops.

When two genes are located on the same chromosome, the chance of a crossover producing recombination between the genes is related to the distance between the two genes. Thus, the use of recombination frequencies has been used to develop linkage maps or genetic maps. However, it is important to note that recombination frequency tends to underestimate the distance between two linked genes. This is because as the two genes are located further apart, the chance of double or even number of crossovers between them also increases. Double or even number of crossovers between the two genes results in them being cosegregated to the same gamete, yielding a parental progeny instead of the expected recombinant progeny. The diploid nature of chromosomes allows for genes on different chromosomes to assort independently during sexual reproduction, recombining to form new combinations of genes. Genes on the same chromosome would theoretically never recombine, however, were it not for the process of chromosomal crossover. During crossover, chromosomes exchange stretches of DNA, effectively shuffling the gene alleles between the chromosomes. This process of chromosomal crossover generally occurs during meiosis, a series of cell divisions that creates haploid cells.

The probability of chromosomal crossover occurring between two given points on the chromosome is related to the distance between the points. For an arbitrarily long distance, the probability of crossover is high enough that the inheritance of the genes is effectively uncorrelated. For genes that are closer together, however, the lower probability of crossover means that the genes demonstrate genetic linkage—alleles for the two genes tend to be inherited together. The amounts of linkage between a series of genes can be combined to form a linear linkage map that roughly describes the arrangement of the genes along the chromosome.

Heredity

Heredity is the passing of traits to offspring (from its parent or ancestors). This is the process by which an offspring cell or organism acquires or becomes predisposed to the characteristics of its parent cell or organism. Through heredity, variations exhibited by individuals can accumulate and cause a species to evolve. The study of heredity in biology is called genetics, which includes the field of epigenetics. Ancients had a variety of ideas about heredity: Theophrastus proposed that male flowers caused female flowers to ripen; Hippocrates speculated that "seeds" were produced by various body parts and transmitted to offspring at the time of conception, and Aristotle thought that male and female semen mixed at conception. Aeschylus, in 458 BC, proposed the male as the parent, with the female as a "nurse for the young life sown within her." Various hereditary mechanisms were envisaged without being properly tested or quantified. These included blending inheritance and the inheritance of acquired traits. Nevertheless, people were able to develop domestic breeds of animals as well as crops through artificial selection. The inheritance of acquired traits also formed a part of early Lamarckian ideas on evolution.

In the 9th century AD, the Afro-Arab writer Al-Jahiz considered the effects of the environment on the likelihood of an animal to survive, and first described the struggle for existence. His ideas on the struggle for existence in the *Book of Animals* have been summarized as follows:

> *"Animals engage in a struggle for existence; for resources, to avoid being eaten and to breed. Environmental factors influence*

organisms to develop new characteristics to ensure survival, thus transforming into new species. Animals that survive to breed can pass on their successful characteristics to offspring."

In 1000 AD, the Arab physician, Abu al-Qasim al-Zahrawi (known as Albucasis in the West), wrote the first clear description of haemophilia, a hereditary genetic disorder, in his *Al-Tasrif*. In this work, he wrote of an Andalusian family whose males died of bleeding after minor injuries. During the 18th century, Dutch microscopist Antonie van Leeuwenhoek (1632–1723) discovered "animalcules" in the sperm of humans and other animals.

Some scientists speculated they saw a "little man" (homunculus) inside each sperm. These scientists formed a school of thought known as the "spermists." They contended the only contributions of the female to the next generation were the womb in which the homunculus grew, and prenatal influences of the womb. An opposing school of thought, the ovists, believed that the future human was in the egg, and that sperm merely stimulated the growth of the egg. Ovists thought women carried eggs containing boy and girl children, and that the gender of the offspring was determined well before conception.

Pangenesis was an idea that males and females formed "pangenes" in every organ. These pangenes subsequently moved through their blood to the genitals and then to the children. The concept originated with the ancient Greeks, and influenced biology until as recently as a century ago. The terms "blood relative," "bloodline," "full-blooded," and "royal blood" are relics of pangenesis. Francis Galton, Charles Darwin's cousin, experimentally tested and disproved pangenesis during the 1870s.

Types of Heredity

An allele is said to be dominant if it is always expressed in the appearance of an organism (phenotype). For example, in peas the allele for green pods, G, is dominant to that for yellow pods, g. Since the allele for green pods is dominant, pea plants with the pair of alleles GG (homozygote) or Gg (heterozygote) will have green pods. The allele for yellow pods is recessive. The effects of this allele are only seen when it is present in both chromosomes, gg (homozygote).

The description of a mode of biological inheritance consists of three main categories:

1. Number of involved loci;
    - Monogenetic (also called "simple") – one locus
    - Oligogenetic – few loci
    - Polygenetic – many loci.

2. Involved chromosomes;
   - Autosomal – loci are not situated on a sex chromosome
   - Gonosomal – loci are situated on a sex chromosome
     - X-chromosomal – loci are situated on the X chromosome (the more common case)
     - Y-chromosomal – loci are situated on the Y chromosome
   - Mitochondrial – loci are situated on the mitochondrial DNA.
3. Correlation genotype–phenotype;
   - Dominant
   - Intermediate (also called "codominant")
   - Recessive.

These three categories are part of every exact description of a mode of inheritance in the above order. Additionally, more specifications may be added as follows:

4. Coincidental and environmental interactions;
   - Penetrance
     - Complete
     - Incomplete (percentual number)
   - Expressivity
     - Invariable
     - Variable
   - Heritability (in polygenetic and sometimes also in oligogenetic modes of inheritance)
   - Maternal or paternal imprinting phenomena.
5. Sex-linked interactions;
   - Sex-linked inheritance (gonosomal loci)
   - Sex-limited phenotype expression (e.g., cryptorchism)
   - Inheritance through the maternal line (in case of mitochondrial DNA loci)
   - Inheritance through the paternal line (in case of Y-chromosomal loci).
6. Locus–locus interactions;
   - Epistasis with other loci (e.g., overdominance)
   - Gene coupling with other loci

- Homozygotous lethal factors
- Semi-lethal factors.

Determination and description of a mode of inheritance is primarily achieved through statistical analysis of pedigree data. In case the involved loci are known, methods of molecular genetics can also be employed. When Charles Darwin proposed his theory of evolution in 1859, one of its major problems was the lack of an underlying mechanism for heredity. Darwin believed in a mix of blending inheritance and the inheritance of acquired traits (pangenesis). Blending inheritance would lead to uniformity across populations in only a few generations and thus would remove variation from a population on which natural selection could act.

This led to Darwin adopting some Lamarckian ideas in later editions of *On the Origin of Species* and his later biological works. Darwin's primary approach to heredity was to outline how it appeared to work (noticing that traits could be inherited which were not expressed explicitly in the parent at the time of reproduction, that certain traits could be sex-linked, etc.) rather than suggesting mechanisms. Darwin's initial model of heredity was adopted by, and then heavily modified by, his cousin Francis Galton, who laid the framework for the biometric school of heredity. Galton rejected the aspects of Darwin's pangenesis model which relied on acquired traits.

The inheritance of acquired traits was shown to have little basis in the 1880s when August Weismann cut the tails off many generations of mice and found that their offspring continued to develop tails. The idea of particulate inheritance of genes can be attributed to the Moravian monk Gregor Mendel who published his work on pea plants in 1865. However, his work was not widely known and was rediscovered in 1901. It was initially assumed the Mendelian inheritance only accounted for large (qualitative) differences, such as those seen by Mendel in his pea plants—and the idea of additive effect of (quantitative) genes was not realised until R.A. Fisher's (1918) paper, "The Correlation Between Relatives on the Supposition of Mendelian Inheritance."

In the 1930s, work by Fisher and others resulted in a combination of Mendelian and biometric schools into the modern evolutionary synthesis. The modern synthesis bridged the gap between experimental geneticists and naturalists; and between both and palaeontologists, stating that:

1. All evolutionary phenomena can be explained in a way consistent with known genetic mechanisms and the observational evidence of naturalists.
2. Evolution is gradual: small genetic changes, recombination ordered by natural selection. Discontinuities amongst species (or other taxa) are explained as originating gradually through geographical separation and extinction (not saltation).

3. Selection is overwhelmingly the main mechanism of change; even slight advantages are important when continued. The object of selection is the phenotype in its surrounding environment. The role of genetic drift is equivocal; though strongly supported initially by Dobzhansky, it was downgraded later as results from ecological genetics were obtained.
4. The primacy of population thinking: the genetic diversity carried in natural populations is a key factor in evolution. The strength of natural selection in the wild was greater than expected; the effect of ecological factors such as niche occupation and the significance of barriers to gene flow are all important.
5. In palaeontology, the ability to explain historical observations by extrapolation from micro to macro-evolution is proposed. Historical contingency means explanations at different levels may exist. Gradualism does not mean constant rate of change.

The idea that speciation occurs after populations are reproductively isolated has been much debated. In plants, polyploidy must be included in any view of speciation. Formulations such as 'evolution consists primarily of changes in the frequencies of alleles between one generation and another' were proposed rather later. The traditional view is that developmental biology ('evo-devo') played little part in the synthesis, but an account of Gavin de Beer's work by Stephen Jay Gould suggests he may be an exception. Almost all aspects of the synthesis have been challenged at times, with varying degrees of success. There is no doubt, however, that the synthesis was a great landmark in evolutionary biology. It cleared up many confusions, and was directly responsible for stimulating a great deal of research in the post-World War II era. Trofim Lysenko however caused a backlash of what is now called Lysenkoism in the Soviet Union when he emphasised Lamarckian ideas on the inheritance of acquired traits. This movement affected agricultural research and led to food shortages in the 1960s and seriously affected the USSR.

Gene Interaction

Gene–environment interaction (or genotype–environment interaction or GxE) is the phenotypic effect of interactions between genes and the environment. Gene–environment interaction is exploited by plant and animal breeders to benefit agriculture. For example, plants can be bred to have tolerance for specific environments, such as high or low water availability. The way that trait expression varies across a range of environments for a given genotype is called its norm of reaction. In genetic epidemiology it is frequently observed that diseases cluster in families, but family members may not inherit disease as such. Often, they inherit sensitivity to the effects of various environmental risk factors. Individuals may be differently affected by exposure to the same environment in medically significant ways. For example, sunlight exposure has a much stronger influence on skin cancer risk in fair-skinned humans than in individuals with an inherited tendency to darker skin.

Naive nature versus nurture debates assume that variation in a given trait is primarily due to either genetic variability or exposure to environmental experiences. The current scientific view is that neither genetics nor environment are solely responsible for producing individual variation, and that virtually all traits show gene–environment interaction. Evidence of statistical interaction between genetic and environmental risk factors is often used as evidence for the existence of an underlying mechanistic interaction. In some combinations of environments and genotypic ranges, heritability can be 100% even while group differences are completely environmental. For heritability to be 100%, random variation in expression must not occur.

1. A classic example of gene–environment interaction is Tryon's artificial selection experiment on maze-running ability in rats. Tryon produced a remarkable difference in maze running ability in two selected lines after seven generations of selecting "bright" and "dull" lines by breeding the best and worst maze running rats with others of similar abilities. The difference between these lines was clearly genetic since offspring of the two lines, raised under identical typical lab conditions, performed too differently. This difference disappeared in a single generation, if those rats were raised in an enriched environment with more objects to explore and more social interaction. This result shows that maze running ability is the product of a gene-by-environment interaction; the genetic effect is only seen under some environmental conditions.
2. Seven genetically distinct yarrow plants were collected and three cuttings taken from each plant. One cutting of each genotype was planted at low, medium, and high elevations, respectively. When the plants matured, no one genotype grew best at all altitudes, and at each altitude the seven genotypes fared differently. For example, one genotype grew the tallest at the medium elevation but attained only middling height at the other two elevations. The best growers at low and high elevation grew poorly at medium elevation. The medium altitude produced the worst overall results, but still yielded one tall and two medium-tall samples. Altitude had an effect on each genotype, but not to the same degree nor in the same way.
3. Phenylketonuria (PKU) is a human genetic condition caused by mutations to a gene coding for a particular liver enzyme. In the absence of this enzyme, an amino acid known as phenylalanine does not get converted into the next amino acid in a biochemical pathway, and therefore too much phenylalanine passes into the blood and other

tissues. This disturbs brain development leading to mental retardation and other problems. PKU affects approximately 1 out of every 15,000 infants in the U.S. However, most affected infants do not grow up impaired because of a standard screening program used in the U.S. and other industrialized societies. Newborns found to have high levels of phenylalanine in their blood can be put on a special, phenylalanine-free diet. If they are put on this diet right away and stay on it, these children avoid the severe effects of PKU.

4. A functional polymorphism in the monoamine oxidase A (MAOA) gene promoter can moderate the association between early life trauma and increased risk for violence and antisocial behaviour. Low MAOA activity is a significant risk factor for aggressive and antisocial behaviour in adults who report victimization as children. Persons who were abused as children but have a genotype conferring high levels of MAOA expression are less likely to develop symptoms of antisocial behaviour. These findings must be interpreted with caution, however, because gene association studies on complex traits are notorious for being very difficult to confirm.

- Doctors are interested in knowing whether disease can be prevented by reducing exposure to environmental risks. Gene–environment interaction means that some people carry genetic factors that confer susceptibility or resistance to a certain disorder in a particular environment. It has been argued that there may be significant public health benefits in using genetic information to stratify the allocation of environmental interventions that prevent disease, although this viewpoint is not universally held.
- Pharmacogenetics is the study of genetic variation that causes people to respond differently to drugs. The clinical importance of pharmacogenetics comes from the possibility that drug treatment can be made safer and more effective when the patient's genotype is known. Pharmacogenetic studies can be considered studies of gene–environment interaction, with drug treatment as the environmental variable.

Timing of Environmental Effects

The popular description of an animal being "born that way" does not necessarily discriminate genetic from environmental effects. In viviparous animals, such as humans, environmental influences may act during either pre- or post-natal development; similarly environmental influences may act before and after hatching to affect development in oviparous animals. Environmental influences in utero may be as strong and lasting as genetic or post-natal environmental influence.

Making Sense of the Complex

It is now becoming possible to identify gene relationships, networks, and epistatic interactions on a systems level. Today, high-throughput experimental tools are available to measure molecular and biochemical data. For example, DNA microarrays allow scientists to gather hundreds of thousands of data points from cells, with transcription level used as the measured phenotype. Then, computational and bioinformatics methods can be used to sift and sort though the massive amounts of biological data to search for epistatic interactions. Once we identify and understand epistatic relationships using techniques such as these, we can apply this knowledge to better diagnose and treat complex diseases.

Genome

In modern molecular biology and genetics, the genome is the entirety of an organism's hereditary information. It is encoded either in DNA or, for many types of virus, in RNA. The genome includes both the genes and the non-coding sequences of the DNA.

Origin of Term

The term was adapted in 1920 by Hans Winkler, Professor of Botany at the University of Hamburg, Germany. In Greek, the word *genome* means I become, I am born, to come into being. The Oxford English Dictionary suggests the name to be a blend of the words *gene* and *chromosome*. A few related-*ome* words already existed, such as *biome* and *rhizome*, forming a vocabulary into which *genome* fits systematically.

Overview

Some organisms have multiple copies of chromosomes, diploid, triploid, tetraploid and so on. In classical genetics, in a sexually reproducing organism (typically eukarya) the gamete has half of the number of chromosome of the somatic cell and the genome is a full set of chromosomes in a gamete. In haploid organisms, including cells of bacteria, archaea, and in organelles including mitochondria and chloroplasts, or viruses, that similarly contain genes, the single or set of circular and/or linear chains of DNA (or RNA for some viruses), likewise constitute the *genome*. The term genome can be applied specifically to mean that stored on a complete set of *nuclear DNA* (i.e., the "nuclear genome") but can also be applied to that stored within organelles that contain their own DNA, as with the "mitochondrial genome" or the "chloroplast genome". Additionally, the genome can comprise nonchromosomal genetic elements such as viruses, plasmids, and transposable elements. When people say that the genome of a sexually reproducing species has been "sequenced", typically they are referring to a determination of the sequences of one set of autosomes and one of each type of sex chromosome, which together represent both of the possible sexes. Even in species that exist in only one sex, what is described as "a genome sequence" may be a composite read from the chromosomes of various individuals. In general use, the phrase "genetic makeup" is sometimes used conversationally to mean the genome of a particular individual or organism.

The study of the global properties of genomes of related organisms is usually referred to as genomics, which distinguishes it from genetics which generally studies the properties of single genes or groups of genes. Both the number of base pairs and the number of genes vary widely from one species to another, and there is only a rough correlation between the two (an observation known as the C-value paradox). At present, the highest known number of genes is around 60,000, for the protozoan causing trichomoniasis, almost three times as many as in the human genome.

An analogy to the human genome stored on DNA is that of instructions stored in a library:

- The library would contain 46 books (chromosomes)
- The books range in size from 400 to 3340 pages (genes)
- which is 48 to 250 million letters (A,C,G,T) per book.
- Hence the library contains over six billion letters total;
- The library fits into a cell nucleus the size of a pinpoint;
- A copy of the library (all 46 books) is contained in almost every cell of our body.

Types

Most biological entities that are more complex than a virus sometimes or always carry additional genetic material besides that which resides in their chromosomes. In some contexts, such as sequencing the genome of a pathogenic microbe, "genome" is meant to include information stored on this auxiliary material, which is carried in plasmids. In such circumstances then, "genome" describes all of the genes and information on non-coding DNA that have the potential to be present.

In eukaryotes such as plants, protozoa and animals, however, "genome" carries the typical connotation of only information on chromosomal DNA. So although these organisms contain chloroplasts and/or mitochondria that have their own DNA, the genetic information contained by DNA within these organelles is not considered part of the genome. In fact, mitochondria are sometimes said to have their own genome often referred to as the "mitochondrial genome". The DNA found within the chloroplast may be referred to as the "plastome".

Genomes and Genetic Variation

Note that a genome does not capture the genetic diversity or the genetic polymorphism of a species. For example, the human genome sequence in principle could be determined from just half the information on the DNA of one cell from one individual. To learn what variations in genetic information underlie particular traits or diseases requires comparisons across individuals. This point explains the common usage of "genome" (which parallels a common usage of "gene") to refer not to the information in any particular DNA sequence, but to a whole family of sequences that share a biological context.

Although this concept may seem counter intuitive, it is the same concept that says there is no particular shape that is the shape of a cheetah. Cheetahs vary, and so do the sequences of their genomes. Yet both the individual animals and their sequences share commonalities, so one can learn something about cheetahs and "cheetah-ness" from a single example of either.

Sequencing and Mapping

The Human Genome Project was organized to map and to sequence the human genome. Other genome projects include mouse, rice, the plant *Arabidopsis thaliana*, the puffer fish, bacteria like E. coli, etc. In 1976, Walter Fiers at the University of Ghent (Belgium) was the first to establish the complete nucleotide sequence of a viral RNA-genome (bacteriophage MS2). The first DNA-genome project to be completed was the Phage Ö-X174, with only 5386 base pairs, which was sequenced by Fred Sanger in 1977. The first bacterial genome to be completed was that of Haemophilus influenzae, completed by a team at The Institute for Genomic Research in 1995.

The development of new technologies has dramatically decreased the difficulty and cost of sequencing, and the number of complete genome sequences is rising rapidly. Among many genome database sites, the one maintained by the US National Institutes of Health is inclusive.

These new technologies open up the prospect of personal genome sequencing as an important diagnostic tool. A major step towards that goal was the May 2007 *New York Times* announcement that the full genome of DNA pioneer James D. Watson was deciphered.

Whereas a genome sequence lists the order of every DNA base in a genome, a genome map identifies the landmarks. A genome map is less detailed than a genome sequence and aids in navigating around the genome.

Genome Evolution

Genomes are more than the sum of an organism's genes and have traits that may be measured and studied without reference to the details of any particular genes and their products. Researchers compare traits such as *chromosome number* (karyotype), genome size, gene order, codon usage bias, and GC-content to determine what mechanisms could have produced the great variety of genomes that exist today.

Duplications play a major role in shaping the genome. Duplications may range from extension of short tandem repeats, to duplication of a cluster of genes, and all the way to duplications of entire chromosomes or even entire genomes. Such duplications are probably fundamental to the creation of genetic novelty.

Horizontal gene transfer is invoked to explain how there is often extreme similarity between small portions of the genomes of two organisms that are otherwise very distantly related. Horizontal gene transfer seems to be common among many microbes.

Also, eukaryotic cells seem to have experienced a transfer of some genetic material from their chloroplast and mitochondrial genomes to their nuclear chromosomes.

3

Visualising Human Chromosomes

Human Genome

The human genome is the genome of *Homo sapiens*, which is stored on 23 chromosome pairs. 22 of these are autosomal chromosome pairs, while the remaining pair is sex-determining. The haploid human genome occupies a total of just over 3 billion DNA base pairs. The Human Genome Project (HGP) produced a reference sequence of the euchromatic human genome, which is used worldwide in biomedical sciences. The haploid human genome contains ca. 23,000 protein-coding genes, far fewer than had been expected before its sequencing. In fact, only about 1.5% of the genome codes for proteins, while the rest consists of non-coding RNA genes, regulatory sequences, introns, and noncoding DNA (once known as "junk DNA").

Features

Genes

There are estimated to be between 20,000 and 25,000 human protein-coding genes. The estimate of the number of human genes has been repeatedly revised down as genome sequence quality and gene finding methods have improved. Earlier predictions estimated that human cells have as much as 200,000 genes. Surprisingly, the number of human genes seems to be less than a factor of two greater than that of many much simpler organisms, such as the roundworm and the fruit fly. However, human cells make extensive use of alternative splicing to produce several different proteins from a single gene, and the human proteome is thought to be much larger than those of the aforementioned organisms. Besides, most human genes have multiple exons, and human introns are frequently much longer than the flanking exons. Human genes are distributed unevenly across the chromosomes. Each chromosome contains various gene-rich and gene-poor regions, which seem to be correlated with chromosome bands and GC-content.

The significance of these nonrandom patterns of gene density is not well understood. In addition to protein coding genes, the human genome contains thousands of RNA genes, including tRNA, ribosomal RNA, microRNA, and other non-coding RNA genes.

Regulatory Sequences

The human genome has many different regulatory sequences which are crucial to controlling gene expression. These are typically short sequences that appear near or within genes. A systematic understanding of these regulatory sequences and how they together act as a gene regulatory network is only beginning to emerge from computational, high-throughput expression and comparative genomics studies. Some types of non-coding DNA are genetic "switches" that do not encode proteins, but do regulate when and where genes are expressed. Identification of regulatory sequences relies in part on evolutionary conservation. The evolutionary branch between the primates and mouse, for example, occurred 70–90 million years ago. So computer comparisons of gene sequences that identify conserved non-coding sequences will be an indication of their importance in duties such as gene regulation.

Another comparative genomic approach to locating regulatory sequences in humans is the gene sequencing of the puffer fish. These vertebrates have essentially the same genes and regulatory gene sequences as humans, but with only one-eighth the noncoding DNA. The compact DNA sequence of the puffer fish makes it much easier to locate the regulatory genes.

Other DNA

Protein-coding sequences (specifically, coding exons) comprise less than 1.5% of the human genome. Aside from genes and known regulatory sequences, the human genome contains vast regions of DNA the function of which, if any, remains unknown. These regions in fact comprise the vast majority, by some estimates 97%, of the human genome size.

Repeat Elements

- Tandem repeats;
  - Satellite DNA
  - Minisatellite
  - Microsatellite.

- Interspersed repeats;
  - SINEs
  - LINEs.

Transposons

- Retrotransposons;
  - [[Retrotransposon#LTR retrotransposons | LTR]]
    - Ty1-copia
    - Ty3-gypsy.
  - Non-LTR
    - SINEs
    - LINEs
- DNA Transposons.

Noncoding DNA

There is also a large amount of sequence that does not fall under any known classification. Much of this sequence may be an evolutionary artifact that serves no present-day purpose, and these regions are collectively referred to as noncoding DNA. These regions were once referred to as "junk" DNA; however, there are a variety of emerging indications that many sequences within are likely to function in ways that are not fully understood.

Recent experiments using microarrays have revealed that a substantial fraction of non-genic DNA is in fact transcribed into RNA, which leads to the possibility that the resulting transcripts may have some unknown function. Also, the evolutionary conservation across the mammalian genomes of much more sequence than can be explained by protein-coding regions indicates that many, and perhaps most, functional elements in the genome remain unknown.

The investigation of the vast quantity of sequence information in the human genome whose function remains unknown is currently a major avenue of scientific inquiry. Meanwhile, considering the global genome DNA information as a whole could provide new ways to understand a possible global level function of non coding DNA.

Information Content

The 2.9 billion base pairs of the haploid human genome correspond to about 691.4 megabytes of data, since every base pair can be coded by 2 bits. The entropy rate of the genome differs significantly between coding and non-coding sequences. It is close to the maximum of 2 bits per base pair for the coding sequences (about 45 million base pairs), but less for the non-coding parts. It ranges between 1.5 and 1.9 bits per base pair for the individual chromosome, except for the Y chromosome, which has an entropy rate below 0.9 bits per base pair.

Sequencing

DNA sequencing determines the order of the nucleotide bases in a genome.

Composite

The Human Genome Project and a parallel project by Celera Genomics each produced and published a haploid human genome sequence, both of which were a composite of the DNA sequence of several individuals.

Personal

A personal genome sequence is a complete sequencing of the chemical base pairs that make up the DNA of a single person. Because medical treatments have different effects on different people because of genetic variations such as single-nucleotide polymorphisms (SNPs), the analysis of personal genomes may lead to personalized medical treatment based on individual genotypes. The completion of the fifth such map was announced in December 2008. The genome mapped was that of a Korean researcher Seong-Jin Kim. Genome maps had previously been completed for Craig Venter of the U.S. in 2007, James Watson of the U.S. in April 2008, and Yang Huanming of China in November 2008 and Dan Stoicescu in January 2008.

Personal genomes had not been sequenced in the Human Genome Project to protect the identity of volunteers who provided DNA samples. That sequence was derived from the DNA of several volunteers from a diverse population. Another distinction is that the HGP sequence is haploid, however, the sequence maps for Venter and Watson for example are diploid, representing both sets of chromosomes. Kim's genome had 1.58 million SNPs that had never been reported before and indicates that six out of 10,000 DNA bases are unique to Koreans. Kim's sequence map can be used to assist in building a standard Korean genome, which can then be used to compare the genomes of other Korean individuals for personalized medical treatments.

Mapping

Whereas a genome sequence lists the order of every DNA base in a genome, a genome map identifies the landmarks. A genome map is less detailed than a genome sequence and aids in navigating around the genome.

Variation

An example of a variation map is the HapMap being developed by the International HapMap Project. The HapMap is a haplotype map of the human genome, "which will describe the common patterns of human DNA sequence variation." It catalogs the patterns of small-scale variations in the genome that involve single DNA letters, or bases. Researchers published the first sequence-based map of large-scale structural variation across the human genome in the journal *Nature* in May 2008. Large-scale structural variations are differences in the genome among people that range from a few thousand to a few million DNA bases; some are gains or losses of stretches of genome sequence and others appear as re-arrangements of stretches of sequence. These variations include differences in the number of copies individuals have of a particular gene, deletions, translocations and inversions.

Variation

Most studies of human genetic variation have focused on single-nucleotide polymorphisms (SNPs), which are substitutions in individual bases along a chromosome. Most analyses estimate that SNPs occur on average somewhere between every 1 in 100 and 1 in 300 base pairs in the euchromatic human genome, although they do not occur at a uniform density. Thus follows the popular statement that "we are all, regardless of race, genetically 99.9% the same", although this would be somewhat qualified by most geneticists. For example, a much larger fraction of the genome is now thought to be involved in copy number variation. A large-scale collaborative effort to catalogue SNP variations in the human genome is being undertaken by the International HapMap Project.

The genomic loci and length of certain types of small repetitive sequences are highly variable from person to person, which is the basis of DNA fingerprinting and DNA paternity testing technologies. The heterochromatic portions of the human genome, which total several hundred million base pairs, are also thought to be quite variable within the human population (they are so repetitive and so long that they cannot be accurately sequenced with current technology). These regions contain few genes, and it is unclear whether any significant phenotypic effect results from typical variation in repeats or heterochromatin. Most gross genomic mutations in gamete germ cells probably result in inviable embryos; however, a number of human diseases are related to large-scale genomic abnormalities. Down syndrome, Turner Syndrome, and a number of other diseases result from nondisjunction of entire chromosomes. Cancer cells frequently have aneuploidy of chromosomes and chromosome arms, although a cause and effect relationship between aneuploidy and cancer has not been established.

Genetic Disorders

Most aspects of human biology involve both genetic (inherited) and non-genetic (environmental) factors. Some inherited variation influences aspects of our biology that are not medical in nature (height, eye colour, ability to taste or smell certain compounds, etc). Moreover, some genetic disorders only cause disease in combination with the appropriate environmental factors (such as diet). With these caveats, genetic disorders may be described as clinically defined diseases caused by genomic DNA sequence variation.

In the most straightforward cases, the disorder can be associated with variation in a single gene. For example, cystic fibrosis is caused by mutations in the CFTR gene, and is the most common recessive disorder in caucasian populations with over 1,300 different mutations known. Disease-causing mutations in specific genes are usually severe in terms of gene function, and are fortunately rare, thus genetic disorders are similarly individually rare. However, since there are many genes that can vary to cause genetic disorders, in aggregate they comprise a significant component of known medical conditions, especially in pediatric medicine. Molecularly characterized genetic disorders are those for which the underlying causal gene has been identified, currently there are approximately 2,200 such disorders annotated in the OMIM database.

Studies of genetic disorders are often performed by means of family-based studies. In some instances population based approaches are employed, particularly in the case of so-called founder populations such as those in Finland, French-Canada, Utah, Sardinia, etc. Diagnosis and treatment of genetic disorders are usually performed by a geneticist-physician trained in clinical/medical genetics. The results of the Human Genome Project are likely to provide increased availability of genetic testing for gene-related disorders, and eventually improved treatment. Parents can be screened for hereditary conditions and counselled on the consequences, the probability it will be inherited, and how to avoid or ameliorate it in their offspring. As noted above, there are many different kinds of DNA sequence variation, ranging from complete extra or missing chromosomes down to single nucleotide changes.

It is generally presumed that much naturally occurring genetic variation in human populations is phenotypically neutral, i.e. has little or no detectable effect on the physiology of the individual (although there may be fractional differences in fitness defined over evolutionary time frames). Genetic disorders can be caused by any or all known types of sequence variation. To molecularly characterize a new genetic disorder, it is necessary to establish a causal link between a particular genomic sequence variant and the clinical disease under investigation. Such studies constitute the realm of human molecular genetics. With the advent of the Human Genome and International HapMap Project, it has become feasible to explore subtle genetic influences on many common disease conditions such as diabetes, asthma, migraine, schizophrenia, etc. Although some causal links have been made between genomic sequence variants in particular genes and some of these diseases, often with much publicity in the general media, these are usually not considered to be genetic disorders *per se* as their causes are complex, involving many different genetic and environmental factors. Thus there may be disagreement in particular cases whether a specific medical condition should be termed a genetic disorder.

Evolution

Comparative genomics studies of mammalian genomes suggest that approximately 5% of the human genome has been conserved by evolution since the divergence of extant lineages approximately 200 million years ago, containing the vast majority of genes. Intriguingly, since genes and known regulatory sequences probably comprise less than 2% of the genome, this suggests that there may be more unknown functional sequence than known functional sequence. A smaller, yet substantial, fraction of human genes seem to be shared among most known vertebrates. The published chimpanzee genome differs from that of the human genome by 1.23% in direct sequence comparisons.

Around 20% of this figure is accounted for by variation within each species, leaving only ~1.06% consistent sequence divergence between humans and chimps at shared genes. This nucleotide by nucleotide difference is dwarfed, however, by the portion of each genome that is not shared, including around 6% of functional genes that are unique to either humans or chimps. In other words, the considerable observable differences between humans and chimps may be due as much or more to genome level variation in the number, function and expression of genes rather than DNA sequence changes in shared genes. On average, a typical human protein-coding gene differs from its chimpanzee ortholog by only two amino acid substitutions; nearly one third of human genes have exactly the same protein translation as their chimpanzee orthologs. A major difference between the two genomes is human chromosome 2, which is equivalent to a fusion product of chimpanzee chromosomes 12 and 13 (later renamed to chromosomes 2A and 2B, respectively). Humans have undergone an extraordinary loss of olfactory receptor genes during our recent evolution, which explains our relatively crude sense of smell compared to most other mammals. Evolutionary evidence suggests that the emergence of colour vision in humans and several other primate species has diminished the need for the sense of smell.

Chimpanzee Genome Project

The Chimpanzee Genome Project is an effort to determine the DNA sequence of the Chimpanzee genome. It is expected that by comparing the genomes of humans and other apes, it will be possible to better understand what makes humans distinct from other species. Chimp-human chromosome differences. The major structural difference is that human chromosome 2 (green colour code) was derived from two smaller chromosomes that are found in other great apes. Parts of human chromosome 2 are scattered among parts of several cat and rat chromosomes in these species that are more distantly related to humans (more ancient common ancestors; about 85 million years since the human/rodent common ancestor: Entrez Pubmed 12552136)

Identification of Human Chromsome

The recognition of chromosome bonds has provided new morphologic features for use in the identification of human chromosomes and chromosomal parts. In addition to the telomere, centromere, and arms, special landmarks (*i.e.*, well-defined bands) were selected to subdivides the arms into regions. The regions are designaled 1,2,3, and 4, moving from the centromere towards the telomere. Using the nomenclater established at the Paris conference is 1971, any particular band or segment of a chromosome can be identified according to the following parameters:

- Chromosome number,
- An arm symbol (p=short arm, q=long),
- The region and band numbers.

For example, 6p23 indicates band number 3 in region 2 of the short arm of chromosome 6. With the use of the banding techniques, particularly the G and Q banding some 320 bands per haploid set can be identified at metaphase. However, if mitosis of followed through late prophase and prometaphase, up to 1256 bands may be recognized.

Determination of Sex

The well-known fact that male and female individuals are statistically found in about equal members suggests that sex is determined by a hereditary mechanism There is physiology and cytologic proof that sex is determined as soon as the egg is fertilized and that it depends on the gametes. Among the physiology evidence is the finding that identical twins—which originate from a single zygote—are always of the same sex. Cytologic evidence was first obtained by McClug, who demonstrated that the karyorpe of a cell is composed not only of common chromosomes (*autosomes*) but also of one or more special chromosomes that are distinguished from the autosomes by morphologic characteristics and behaviour: These were called *accessory chromosomes, allosomes, heterochromsomes, of sex chromosomes.*

The majority of organisms have a pair of sex chromsome, which in the course of evolution, have been specialized for sex determination. One of the sexes has a pair of identical sex chromosomes (XX); the other may have a single sex chromosome, which may be unpaired (XO) or paired with a Y chromosome (XY) The XY pair is also called *hetermorphic* because of the different morphology of the chromosomes. In the human the garments are not identical with respect to the sex chromosomes. The male is heterozygous (XY) and produces two types of spermatozoa in similar proportions, *i.e.*, one carries the X chromosome, and the other the Y. the female, being homozygous (XX), produces only one kind of gamete (the ovum). Therefore only two combinations of gametes are possible at fertilization and the result is 50 per cent males and 50 per cent females. This type of sex determination is found in all mammals and in certain insects, such as *Drosophila.* In other vertebrates and certain invertebrates, the female is heterogametic, and the male, homogametic. In some animals the sex chromosomes cannot be distinguished from the other. In such cases the sexdetermining genes are confined to a region in a pair of chromosomes.

In the human the Y chromosome contains an essential testis-determining factor and determines the male sex. The genes involved are situated in a region near the centromere. An individual with XO (lacking the Y chromosome) will resemble a female but will have atrophied ovaries (Turner's syndrome). We have seen earlier that the X and Y chromosomes differ in size and morphology. In the XY pair there is a *homologous region* in which recombination may take place during meiosis and a *differential region* that is unpaired.

Interphase Nuclei

The X chromatin can be bound as a small body in different positions within the nucleus. For example, in nerve cells it may be near the nucleolus in the nucleoplasm, or near the nuclear envelope. In cells of the oral mucosa it is generally attached to the envelope, and in neutrophil leukocytes it may appear as a small rod called the *drumstick.* The study of sex chromatin has a wide field of medical applications and offers the possibility of relating the origin of certain congenital diseases to chromosome anomalies. Among these application is the diagnosis of sex in intersexual states in postnatal and even in fetal life. The frequency with which sex chromation can be detected in the female varies from tissue to tissue. In nervous tissue the frequency may be 85 per cent, whereas in whole mounts of amniotic or chorionic epithelium it may be as high as 96 per cent. In oral smears the frequency may vary between 20 and 50 per cent in normal females.

The differential behaviour of the two X chromosome in the female led Lyon to the so-called *inactive* X *hypothesis* is genetically active; the X undergoing heteropyknosis may be either of maternal or paternal origin, and the decision by which X becomes inactive is taken at random; the inactivation occures early in embryonic life and remains fixed. It is now admitted that only a part of the X chromosome condenses into a Barr corpuscle; This condensed chromosome is said to contain *facultative heterochromatin,* as opposed to the *constitutive heterochromatin* found in other chromosome takes place even in 3X and 4X individuals, in which only one X remains uncondensed. The inactivation starts in the human in the late blastocyst stage on about the sixteenth day of embryonic life; the inactivated X chromosome then remains heterochromatic in the somatic cells. The fact that X inactivation occurs at random has been demonstrated in human diseases linked to the X chromosome. The *Lesch-Nyhansyndrome,* in which a deficiency of one enzyme of the purine metabolism (*i.e.*, hypoxanthine-guanine phosphoribosyl transferase) produces mental retardation and increase uric acid levels, results from a recessive mutation in the X chromsome. When fibroblasts of heterozygous women are cultured in vitro, two types of cell clones are obtained. Half the clones contain the enzyme, whereas the other half lack the enzyme. During the cell cycle the heterochromatic X chromosome is characterized by its late replication. This can be clearly observed by the use of 5-bromodeoxyuridine, followed by the fluorescent dye. Hoechst 33258 or the Giemsa stain.

The above-mentioned fluorescence method for the study of Q banding led to the demonstration of that a large portion of the Y chromosome is heterochromatic and appears, at interphase, as a strongly fluorescent body, the so- called Y *chromatin*. While the X chromatin is 1.0 to 1.2 μm in diameter, the Y chromatin is only about 0.7-μm in diameter and generally is not attached to the nuclear envelope. The number of Y chromation bodies is identical to the number of Y chromosome. For example, individuals with XYY have two Y chromatin bodies. The human Y chromosome consists of two segments, one that does not fluoresce and is genetically active. and a second that fluoresces and is genetically inactive. It is this latter segment that may be polymorphic, showing wide variations in size between individuals. Among mammal, only man and the gorilla share this fluorescent region of the Y chromosome, suggesting that it is of recent evolutionalry origin.

This region corresponds to a highly repetitive (satellite) DNA, which is specific fore the Y chromosome and probably has no genetic function. Although the primary determination of sex is made at fertilization the embryo acquires its definite sex characteristics by a more complex mechanism. An epigenetic factor (*i.e.*, hormonal) may assume control of the genetic determination during development, thereby changing the phenotypic direction of sex. Among vertebrates a condition of bisexuality may exist (*e.g.*, the coexistence of structures of the functional sex, together with primordia of the heterologous sex). For example, male amphibians have a redimentary ovary (Bidder's organ) and vestigial oviducts. In the human embryo until the sixth week the gonads and the primordia of the urogenital tract are identical in males and females. At this time the gonad has already has already been invaded by the primary XX or XY germ cells. At this point, a gene (or set of genes) present in the y chromosome causes the undifferentiated gonad to differentiate into a testis, and the absence of this gene allows the gonad to become an ovary. The development into a testis, and the absence of this gene allows the gonad to become an ovary. The development into a testis starts as soon as the *gonocytes* (*i.e.*, primordial germ cells) from the yolk sac have finished their migration into the *gonadal ridge.* Gonocytes of the male (XY) migrate deeper into the gonadal blastema, and female gonocytes (XX) remain at the periphery, forming a thick cortical layer.

H-Y Antigen

It is known that the Y chromosome directs the organogenesis of the testis by inducing the production of a protein (or membrane and determines, by cell recognition, the formation of the seminiferous tubule. It has been proposed that a Y-linked determining gene of mammals produces that a Y-linked determining gene of mammals produces this *H-T antigen,* which is found in various cell types in males but is absent in females. This antigen is also a histocompatibility factor (H) that is found, for example, in the skin of male mice and that determines the rejection of female skin transplants.

Action of Testosterone

The gonad differentiates into a definite testicle at the seventh week, whereas the female gonad differentiates between the eighth and ninth weeks of development, At the time of differentiation an important epigenetic factor is the production of androgens by somatic cells in the embryonic male gonad. In mammals administration of *testosterone* (the male hormone) to the mother produces in the fetus a shift in the differentiation of XX genitalia to a male type, producting what is called *feminine masculine pseudohermaphroditism.* This hormone acting locally accelerates the development of the testis, whereas in the female the absence of the hormone permits the slower development characteristic of the ovary.

Chromosomal Abnormalities

The study of chromosome abnormalities in man, like that of normal cytogenetics, can be divided into periods before and after the development of the banding techniques. In the earlier period it was found that 0.5 per cent of all newborns have gross chromosomal abnormalities, about half of which are in the sex chromosomes and the rest of which are in the autosomes. These abnormalities generally consist of aneuploidy, such as monosomy or trisomy, Structural aberrations, such as translocation, deficiency, duplication, and other more complex alterations, have also been observed; Diagnosis of chromosomal aberrations can be made even before a child is born. This is done by removing amitotic fluid by puncturing the mother's abdomen, a procedure called *amniocentesis. which* does not entail risks to the fetus. The amniotic fluid contains cells shed from the fetus and which can be cultured.

This technique is useful for karyotyping the babies off mothers with a history of chromosomal abnormality, or children of those who are carriers of a balanced translocation, in whom that risk of conceiving an affected child is high. In addition, this technique permits the *antenatal* determination of the fetal sex, which may be of importance in certain sex-linked diseases. The improved resolution provided by the banding techniques has not only confirmed the previously observed abnormalities but has also led to the discovery of over 30 new chromosomal syndromes and several malignancies defects.

In aneuploidy the genetic message contained in each chromosome is maintained intact. The alteration is quantitative; it resides in a disequilibrium established by the excess (trisomy) or deficit (monosomy) in the amount of genetic material, Such a dosage effect may be dangerous to the organism and may produce severe anatomical and functional anomalies (*i.e.*, malformations). It will be shown later that trisomy produces changes that are characteristic for each chromosome present in excess; however, in all instances of trisomy the tendency is towards involution of the nervous system, resulting in a more or less severe mental defect. Another important consequence of aneuploidy is spontaneous abortion. A large proportion of aborted fetuses show trisomy in one of the larger chromosomes in the karyotype. Aneuploidy of one the larger chromosomes is generally lethal, being more severe than an equivalent alteration in a smaller chromosome. Malformation mental retardation, sterility, and spontaneously abortion operate as strong selective mechanisms tending to eliminate from the general population those individuals carrying deleterious genetic imbalances. Aneuploidy originates by the mechanism called *Non-disjunction.* The immediate cause of non-disjuction is the lagging, during anaphase, of one sister chromatid which during telophase remains in one of the cells together with the other sister chromatid. This change gives rise to a cell line that lacks one chromosome or has one chromosome in excess for that pair (monosomy and trisomy).

Non-disjunction may occur during *meiosis, giving* rise to an aneuploid *ovum. This* when fertilized by a normal spermatozoon, results in a zygote and later an organism in which all cells are aneuploid. In some cases the alteration may be in the male gamete or in both, thus producing more complex types of abnormalities. If non-disjunction occurs in a *mitotic division,* which precedes the formation of germ cells, the effects are similar to those of *meiotic non-disjunction.* If mitotic non-disjunction occurs during embryonic development, a *mosaic* of different cell lines is produced. We mentioned earlier the various structural aberrations that are found in human chromosomes (*i.e.*, translocation, deficiency, and duplication). The most common of these is the so-called *reciprocal translocation,* in which chromosome segments are exchanged between non-homologous chromosomes. Some of these translocations may result in a chromosome having a length and configuration that may be easily identified. However, in other cases the apparent morphology changes very little. Banding techniques are of fundamental importance in detecting such cases. The most important aberrations or as follows.

Klinefelter's Syndrome

Most affected individuals are practically normal except for minor phenotypic anomalies. They have small testes, frequent gynecomastia (enlarged breasts), tendency to tallness, obesity and underdevelopment of secondary sex characteristics. Spermatogenesis does not occur, thereby resulting in complete infertility.

These individuals have a positive sex chromatin and 47 chromosomes (44 autosomes + XXY). Males with 48 chromosome (44 autosome+XXXY) and two Barr corpuscles have also been described. These individuals have features of klinefelter's syndrome and are mentally retarded. Persons with 49 chromosomes (44 autosomes+XXXXY) have also been reported. They display extensive skeletal anomalies, extreme hypogenitalism, and severe mental deficiency. These persons have three X chromation bodies.

XYY Syndrome

Males having two Y chromosomes have been identified in the past in maximum security institutions. It was proposed that such individuals had a strong tendency to ward anti-social behaviour and aggression. More recently, XYY individuals have been found in the normal population in a proportion of 1 in `650 male infants, suggesting that the correlation with violence may not be a strong as was previously thought.

Gonadal Dysgenesis

Patients with Turner's syndrome usually have a female appearance with short stature, webbed neck (folds of skin extended from the mastoid to the shoulders) and generally infantile internal sexual organs. The ovary does not develop and shows complete absence of germ cells. As result of this ovarian dysgenesis, menstruation does not occur, and secondary sexual characteristics do not develop.

The karyotype shows 45 chromosomes (44 autosomes +X), and there is no sexchromation. In this syndrome there is little difference in the gonads up to the third moth of gestation; the ovaries up to the third month of gestation; the ovaries contain approximately the normal number of germ cells (about 2 or 3×10^6 cells) later on there is a rapid atresia of the germ cells, leading to their virtually disappearance after puberty. It is probable that the lack of one X chromosomes determines the progressive ovarian atresia by failure of primordial follicles.

Females with X Polysomy

Triplo-X constitution (47 chromosome; 44 autosomes + XXX) was detected in phenotypically near-normal females. A number of these females are mentally subnormal or psychotic and some of them menstruate. The sex chromatin bodies are found in cells from these women.

A few severely retarded patients with three corpuscles of sex chromatin and 48 chromosomes (44 autosomes + XXXX) have been found. These persons are also called metafermales Mixed Chromosomal Aberrations. Kinefelter's syndrome may be found combined with mongolism. Such a person has 48 chromosomes: 45 autosomes (including a trisomic pair 21) +XXY.

Mosaics

Sometimes chromsomal aberrations are produced during development of the embryo. One interesing example is induced by the loss of one Y chromosomes in the first division of the zygote. This may result in twins, of which one has normal male characters and the other has Tuner's syndrome. Sometimes chromosomal aberrations due to non-disjunction are produced during development of the embryo. These individuals possess different chromosomal complements from cell to cell within the same tissue or between tissues, depending on the embryonic stage at which non-disjunction occurred.

Hermaphroditism

A true hermaphrodite is an individual who has both ovarian and testicular tissue. The two types of gonadal tissue may be separated or in close proximity (ovotestis). Frequently, true hermaphrodites haves a 46, XX chromosome complement and are chromation-positive. Other may have a 46, XY mosaicism has also been detected in this condition.

Mongolism

Before the development of the banding techniques about 12 classic syndromes had been described, the majority of which were caused by an extra chromosome or loss of a chromsome segment. Except for the X monosomy (Turner's syndrome), any other monosomy is believed to be inviable (*i.e.*, leads to death of the embryo). This also holds true for all trisomies other than those of chromosomes 13, 18, or 21.

21-Monosomy

Complete deletion of one of the chromosomes in pair 21 is apparently lethal, but there is a syndrome in which a large part of cone is lacking. Children with this condition have a morphologic aspect which is, to some extent, the opposite of mongolism. The nose is prominent, the distance between eyes shorter than normal, and the ears are large and the muscles contracted, It seems that in trisomy and monosomy of the 21st, the phenotype shifts to one or the other side of normal.

18-Trisomy

In this case the child is small and weak, the head is laterally flattened, and the helix of the ear scarcely developed. The hands are short and the digital imprints are rather simple. These children are very retarded mentally and usually die before one year of age.

18-Monosomy

This is the opposite syndrome is which a partial deletion of one chromosome of the pair occures. The ears are voluminous, the fingers long, and the digital imprints are complex and convoluted.

21-Trisomy Mongolism

Among the most important autosomal aberrations is *mongolism,* which is characterized by multiple malformations, mental retardation, and markedly defective development of the central nervous system. It was discovered that the mongoloid has an extra chromosome. Pair 21 is trisomic instead of normal. This aberration probably originates from non-disjunction of pair 21 during meiosis. The extra chromosome of pair 21 in some cases may become attached to another autosome (translocation), usually to pair 22. The phenotype of a mongoloid is recognizable at birth in most cases. The face of such a patient has a special moon-like aspect, with oblique palpebral fissures, increased separation between the eyes, and a skin fold (epicanthus) at the inner part of the eyes. The nose is flattened the ears and the tongue protrudes.

Mongolism is the most common congenital disease and is present in more than 0.1 per cent of births. Its frequency increases as the mother's age sporadic and in general there is no recurrence in the family. However, in the rarer cases of mongolism by translocation, the disease may affect siblings and may appear in successive generation. Fortunately this "translocation trisomy" represents only 3 to 4 per cent of all call cases of mongolism. In this type there is no change in frequency with the age of the mother, and when the aberration is properly determined by karyotype analysis, the parents should be warned of a repetition of this defect.

13-Trisomy

Multiple and severe body malformations, as well as profound mental deficiency, are characteristic features. The head is small and the eyes are often small, or absent. are-lip, cleft palate, and malformations of the brain are frequent. The internal organs are also severely malformed and in most cases death occurs soon after birth. One of the most remarkable phenotypes associated with a chromosome structural change is the condition known as *cri du chat* syndrome. The affected baby has a strange mewing cry, multiple malfomations and mental retardation associated with partial deletion of the short arm of chromosome 5. Another type of patient showing facial alteration and skeletal and ophthalmologic abnormalities along with profound mental retadation, is the carrier of a deletion of the long arm of chromosome 18.

The use of the banding techniques has permitted the detection of more than 30 new chromosomal syndromes which result from a partial trisomy or deletion from each of almost all the chromosome. Such alterations have been bound in children with mental retardation or minor congenital defects and also in some with normal intelligence and minor anomalies which were earlier considered as deviations of normality. The technical advances have permitted the identification of chromosome defects even when there are few clinical symptoms furthermore, these methods have allowed the study of cases of balanced chromosomal translocations, which, as carriers of this type of syndrome, are important to detect. In such cases transmission can be prevented by proper counselling and prenatal diagnosis.

Chromosomal Aberrations

The development of the banding techniques has also resulted in the demonstration of specific chromosomal defects in certain types of tymors. Here only a few examples will be given. The most characteristic of these is the association of *chronic myelogenous leukemia* with the so-called *Philadelphia chromosome* (Ph 1). This involves a balanced translocation between chromosome 9 and 22. The Ph 1 chromosome is acquired with the disease and is not found in cultured fifbroblasts or in the idential twins of affected patients. In a *retinoblastoma* (*i.e.*, a cancer of the retina) a deletion in the long arm of chromosomes 13 has been identified. Other abnormalities have been observed in lymphomas, meingiomas, and many other tumors. Cancer cells are usually associated with severe chromosomal abnormalities such as polyploidy and aneuploidy.

Starting the Chimpanzee Genome Project

Human and chimpanzee chromosomes are very similar. The primary difference is that humans have one fewer pair of chromosomes than do other great apes. Humans have 23 pairs of chromosomes and other great apes have 24 pairs of chromosomes. In the human evolutionary lineage, two ancestral ape chromosomes fused at their telomeres producing human chromosome 2. There are nine other major chromosomal differences between chimpanzees and humans: chromosome segment inversions on human chromosomes 1, 4, 5, 9, 12, 15, 16, 17, and 18.

After the completion of the Human genome project, a Common Chimpanzee genome project was initiated. In December 2003, a preliminary analysis of 7600 genes shared between the two genomes confirmed that certain genes such as the forkhead-box P2 transcription factor, which is involved in speech development, are different in the human lineage. Several genes involved in hearing were also found to have changed during human evolution, suggesting selection involving human language-related Behaviour. Differences between individual humans and Common Chimpanzees are estimated to be about 10 times the typical difference between pairs of humans.

Draft Genome Sequence of the Common Chimpanzee

Analysis of the genome was published in *Nature* on September 1, 2005, in an article produced by the Chimpanzee Sequencing and Analysis Consortium, a group of scientists which is supported in part by the National Human Genome Research Institute, one of the National Institutes of Health. The article marked the completion of the draft genome sequence. A database now exists containing the genetic differences between human and chimpanzee genes, with about thirty-five million single-nucleotide changes, five million insertion/deletion events, and various chromosomal rearrangements. Gene duplications account for most of the sequence differences between humans and chimps. Single-base-pair substitutions account for about half as much genetic change as does gene duplication.

Typical human and chimp homologs of proteins differ in only an average of two amino acids. About 30 percent of all human proteins are identical in sequence to the corresponding chimp protein. As mentioned above, gene duplications are a major source of differences between human and chimp genetic material, with about 2.7 percent of the genome now representing differences having been produced by gene duplications or deletions during approximately 6 million years since humans and chimps diverged from their common evolutionary ancestor. The comparable variation within human populations is 0.5 percent. About 600 genes have been identified that may have been undergoing strong positive selection in the human and chimp lineages; many of these genes are involved in immune system defence against microbial disease or are targeted receptors of pathogenic microorganisms (example: Glycophorin C and *Plasmodium falciparum*). By comparing human and chimp genes to the genes of other mammals, it has been found that genes coding for transcription factors, such as forkhead-box P2 (FOXP2), have often evolved faster in the human relative to chimp; relatively small changes in these genes may account for the morphological differences between humans and chimps. A set of 348 transcription factor genes code for proteins with an average of about 50 percent more amino acid changes in the human lineage than in the chimp lineage.

Six human chromosomal regions were found that may have been under particularly strong and coordinated selection during the past 250,000 years. These regions contain at least one marker allele that seems unique to the human lineage while the entire chromosomal region shows lower than normal genetic variation. This pattern suggests that one or a few strongly selected genes in the chromosome region may have been preventing the random accumulation of neutral changes in other nearby genes.

One such region on chromosome 7 contains the FOXP2 gene (mentioned above) and this region also includes the Cystic fibrosis transmembrane conductance regulator (CFTR) gene, which is important for ion transport in tissues such as the salt-secreting epithelium of sweat glands. Human mutations in the CFTR gene might be selected for as a way to survive cholera.

Another such region on chromosome 4 may contain elements regulating the expression of a nearby protocadherin gene that may be important for brain development and function. Although changes in expression of genes that are expressed in the brain tend to be less than for other organs (such as liver) on average, gene expression changes in the brain have been more dramatic in the human lineage than in the chimp lineage.

This is consistent with the dramatic divergence of the unique pattern of human brain development seen in the human lineage compared to the ancestral great ape pattern. The protocadherin-beta gene cluster on chromosome 5 also shows evidence of possible positive selection. Results from the human and chimp genome analyses should help in understanding some human diseases. Humans appear to have lost a functional caspase-12 gene, which in other primates codes for an enzyme that may protect against Alzheimer's disease. Figures published in *Nature* on September 1, 2005, in an article produced by the Chimpanzee Sequencing and Analysis Consortium, show that 24% of the chimpanzee genome does not align with the human genome.

There are 3% further alignment gaps, 1.23% SNP differences, and 2.7% copy number variations totaling at least 30% differences between chimpanzee and *Homo sapiens* genomes.

Genes of the Chromosome 2 Fusion Site

The results of the chimpanzee genome project suggest that when ancestral chromosomes 2A and 2B fused to produce human chromosome 2, no genes were lost from the fused ends of 2A and 2B. At the site of fusion, there are approximately 150,000 base pairs of sequence not found in chimpanzee chromosomes 2A and 2B. Additional linked copies of the PGML/FOXD/CBWD genes exist elsewhere in the human genome, particularly near the p end of chromosome 9. This suggests that a copy of these genes may have been added to the end of the ancestral 2A or 2B prior to the fusion event. It remains to be determined if these inserted genes confer a selective advantage.

- PGML. The phosphoglucomutase-like gene of human chromosome 2. This gene is incomplete and may not produce a functional transcript.
- FOXD. The forkhead box D4-like gene is an example of an intronless gene. The function of this gene is not known, but it may code for a transcription control protein.
- CBWD. Cobalamin synthetase is a bacterial enzyme that makes vitamin B_{12}. In the distant past, a common ancestor to mice and apes incorporated a copy of a cobalamin synthetase gene. Humans are unusual in that they have several copies of cobalamin synthetase-like genes, including the one on chromosome 2. It remains to be determined what the function of these human cobalamin synthetase-like genes is. If these genes are involved in vitamin B_{12} metabolism, this could be relevant to human evolution. A major change in human development is greater post-natal brain growth than is observed in other apes. Vitamin B_{12} is important for brain development, and vitamin B_{12} deficiency during brain development results in severe neurological defects in human children.
- CXYorf1-like protein. Several transcripts of unknown function corresponding to this region have been isolated. This region is also present in the closely related chromosome 9p terminal region that contains copies of the PGML/FOXD/CBWD genes.
- Many ribosomal protein L23a pseudogenes are scattered through the human genome.

Mitochondrial Genome

The human mitochondrial genome, while usually not included when referring to the "human genome", is of tremendous interest to geneticists, since it undoubtedly plays a role in mitochondrial disease. It also sheds light on human evolution; for example, analysis of variation in the human mitochondrial genome has led to the postulation of a recent common ancestor for all humans on the maternal line of descent.

Due to the lack of a system for checking for copying errors, Mitochondrial DNA (mtDNA) has a more rapid rate of variation than nuclear DNA. This 20-fold increase in the mutation rate allows mtDNA to be used for more accurate tracing of maternal ancestry. Studies of mtDNA in populations have allowed ancient migration paths to be traced, such as the migration of Native Americans from Siberia or Polynesians from southeastern Asia. It has also been used to show that there is no trace of Neanderthal DNA in the European gene mixture inherited through purely maternal lineage.

Epigenome

Epigenetics are a variety of features of the human genome that transcend its primary DNA sequence, such as chromatin packaging, histone modifications and DNA methylation, and which are important in regulating gene expression, genome replication and other cellular processes. Epigenetic markers strengthen and weaken transcription of certain genes but do not affect the actual sequence of DNA nucleotides.

Craig Venter # Individual Human Genome Sequenced

John Craig Venter (born October 14, 1946) is an American biologist and entrepreneur, most famous for his role in being one of the first to sequence the human genome and for his role in creating the first cell with a synthetic genome in 2010. Venter founded Celera Genomics, The Institute for Genomic Research and the J. Craig Venter Institute, now working at the latter to create synthetic biological organisms and to document genetic diversity in the world's oceans. He was listed on *Time* magazine's 2007 and 2008 Time 100 list of the most influential people in the world. In 2010, The British Magazine New Statesman listed Craig Venter at 14th in the list of "The World's 50 Most Influential Figures 2010".

Early Life

Venter was born in Salt Lake City, Utah. In his youth, Venter did not take his education seriously, preferring to spend his time on the water in boats or surfing. According to his biography, *A Life Decoded*, he was said to never be a terribly engaged student, having Cs and Ds on his eighth-grade report cards. According to *Time Magazine*, it was not always evident that Venter would become a transformative figure, particularly when he was a boy. Although he was against the Vietnam War, Venter was drafted and enlisted in the United States Navy where he worked in the intensive-care ward of a field hospital. Being confronted with wounded, maimed, and dying soldiers on a daily basis instilled in him a desire to study medicine — although he later switched to scientific medical research.

Education

Venter graduated from Mills High School and began his college career at a community college, College of San Mateo in California. He received his B.S. degree in biochemistry in 1972, and his Ph.D. degree in physiology and pharmacology in 1975, both from the University of California, San Diego. At UCSD, he studied under biochemist Nathan O. Kaplan, and married former Ph.D. candidate Barbara Rae. After working as an associate professor, and later as full professor, at the State University of New York at Buffalo, he joined the National Institutes of Health in 1984. In Buffalo, he divorced Dr. Rae-

Venter and married his student, Claire M. Fraser, remaining married to her until 2005.

Discovery

While at the NIH, Venter learned of a technique for rapidly identifying all of the mRNAs present in a cell and began to use it to identify human brain genes. The short cDNA sequence fragments discovered by this method are called expressed sequence tags (ESTs) a name coined by Anthony Kerlavage at The Institute for Genomic Research. The NIH initially led an effort to patent these gene fragments, in which Venter coincidentally and controversially became involved. The NIH later withdrew the patent applications, following public outcry. Subsequent court cases declared that ESTs were not directly patentable.

Human Genome Project

Venter was passionate about the power of genomics to radically transform healthcare. Venter believed that shotgun sequencing was the fastest and most effective way to get useful human genome data. The method was controversial, however, since some geneticists felt it would not be accurate enough for a genome as complicated as that of humans. Frustrated with what Venter viewed as the slow pace of progress in the Human Genome project, and unable to get funds for his ideas, he sought funding from the private sector to fund Celera Genomics. The goal of the company was to sequence the entire human genome and release it into the public domain for non-commercial use in much less time and for much less cost than the public human genome project. The company planned to monetize their work by creating a value-added database of genomic data to which users could subscribe for a fee.

The goal consequently put pressure on the public genome program and spurred several groups to redouble their efforts to produce the full sequence. DNA from five demographically different individuals was used by Celera to generate the sequence of the human genome; one of the individuals was Venter himself. In 2000, Venter and Francis Collins of the National Institutes of Health and U.S. Public Genome Project jointly made the announcement of the mapping of the human genome in 2000, a full three years ahead of the expected end of the Public Genome Program. The announcement was made along with US President Bill Clinton, and U.K. Prime Minister Tony Blair. Venter and Collins thus shared an award for "Biography of the Year" from A&E Network. Celera published the first Human Genome in the journal Science, and was soon followed by a Human Genome Project Publication in Nature. Despite some claims that shotgun sequencing was in some ways less accurate than the clone-by-clone method chosen by the Human Genome Project, the technique became widely accepted by the scientific community and is still the de facto standard used today. Although Celera was originally set to sequence a composite of DNA samples, partway through the sequencing, Venter switched the samples for his own DNA. After contributing to the Human Genome, and its release it into the public domain, Venter was fired by Celera in early 2002. According to his biography, Venter was ready to leave Celera, and was fired due to conflict with the main investor, Tony White, that had existed since day one of the project. Venter writes that his main goal was always to accelerate science and thereby discovery, and he only sought help from the corporate world when he couldn't find funding in the public sector.

Ocean Sampling

The Global Ocean Sampling Expedition (GOS) is an ocean exploration genome project with the goal of assessing the genetic diversity in marine microbial communities and to understand their role in nature's fundamental processes. Begun as a Sargasso Sea pilot sampling project in August 2003, Craig Venter announced the full Expedition on 4 March 2004. The project, which used Craig Venter's personal yacht, *Sorcerer II*, started in Halifax, Canada, circumnavigated the globe and returned to the U.S. in January 2006.

Current Work

Venter is currently the president of the J. Craig Venter Institute, which conducts research in synthetic biology. In June 2005, he co-founded Synthetic Genomics, a firm dedicated to using modified microorganisms to produce clean fuels and biochemicals. In July 2009, ExxonMobil announced a $600 million collaboration with Synthetic Genomics to research and develop next-generation biofuels. Venter is a member of the USA Science and Engineering Festival's Advisory Board.

Media Coverage

Venter has been the subject of articles in several magazines, including *Wired*, *The Economist*, Australian science magazine *Cosmos*, and *The Atlantic*. Additionally, he was featured on *The Colbert Report* on both February 27, 2007, and October 30, 2007. Venter appeared in the "Evolution" episode of the documentary television series *Understanding*. On May 16, 2004, Venter gave the commencement speech at Boston University.

In a 2007 interview with *New Scientist* when asked "Assuming you can make synthetic bacteria, what will you do with them?", Venter replied: Over the next 20 years, synthetic genomics is going to become the standard for making anything. The chemical industry will depend on it. Hopefully, a large part of the energy industry will depend on it. We really need to find an alternative to taking carbon out of the ground, burning it, and putting it into the atmosphere. That is the single biggest contribution I could make.

Furthermore it suggests that one of the main purposes for creating synthetic bacteria would be to reduce the dependence on fossil fuels through bioremediation. On May 10, 2007, Venter was awarded an honorary doctorate from Arizona State University., and on October 24 of the same year, he received an honorary doctorate from Imperial College London. He was on the 2007 Time 100 most influential people in the world list made by Time magazine. In 2007 he also received the Golden Eurydice Award for contributions to Biophilosophy.

On September 4, 2007, a team led by Venter published the first complete (six-billion-letter) genome of an individual human — Venter's own DNA sequence. When on BBC News on October 22, 2007, when asked about his religious view he replied that he thought that a true scientist could not believe in supernatural explanations. On December 4, 2007, Venter gave the Dimbleby lecture for the BBC in London. He outlined his current work and future developments in genetics.

In February 2008, he gave a speech about his current work at the TED conference. Venter delivered the 2008 convocation speech for Faculty of Science honours and specialization students at the University of Alberta. A transcription of the speech is available here. Dr. Venter was featured in Time Magazine's "The Top 10 Everything of 2008" article. Number three in 2008's Top 10 Scientific Discoveries was a piece outlining his work stitching together the 582,000 base pairs necessary to invent the genetic information for a whole new bacterium. Dr. Venter took part in the inaugural San Diego Science Festival and spoke at its press conference on February 26, 2009.

On April 6, 2009, Venter gave a speech at Arizona State University as part of the Origins Symposium. For an episode aired on July 27, 2009, Venter was interviewed on his boat by BBC One for the first episode of TV show Bang Goes the Theory. On May 20, 2010, Venter announced the creation of first self-replicating synthetic bacterial cell. On November 21, 2010 Steve Kroft profiled J. Craig Venter and his research on 60 minutes.

Individual Human Genome Sequenced

On September 4, 2007, a team led by Sam Levy published the first complete (six-billion-letter) genome of an individual human—Venter's own DNA sequence. Some of the sequences in Venter's genome are associated with wet earwax, increased risk of antisocial Behaviour, Alzheimer's and cardiovascular diseases. This publication was especially interesting since it contained a diploid instead of a haploid genome and shows promise for personalized medicine via genotyping. This genome, rather immodestly dubbed HuRef by Levy et al., was a landmark accomplishment and as of mid-2010 is probably the highest quality personal genome sequence yet completed.

The Human Reference Genome Browser is a web application for the navigation and analysis of Venter's recently published genome. The HuRef database consists of approximately 32 million DNA reads sequenced using microfluidic Sanger sequencing, assembled into 4,528 scaffolds and 4.1 million DNA variations identified by genome analysis. These variants include single-nucleotide polymorphisms (SNPs), block substitutions, short and large indels, and structural variations like insertions, deletions, inversions and copy number changes.

The browser enables scientists to navigate the HuRef genome assembly and sequence variations, and to compare it with the NCBI human build 36 assembly in the context of the NCBI and Ensembl annotations. The browser provides a comparative view between NCBI and HuRef consensus sequences, the sequence multi-alignment of the HuRef assembly, Ensembl and dbSNP annotations, HuRef variants, and the underlying variant evidence and functional analysis. The interface also represents the haplotype blocks from which diploid genome sequence can be inferred and the relation of variants to gene annotations. The display of variants and gene annotations are linked to external public resources including dbSNP, Ensembl, Online Mendelian Inheritance in Man (OMIM) and Gene Ontology (GO). Users can search the HuRef genome using HUGO gene names, Ensembl and dbSNP identifiers, HuRef contig or scaffold locations, or NCBI chromosome locations. Users can then easily and quickly browse any genomic region via the simple and intuitive pan and zoom controls; furthermore, data relevant to specific loci can be exported for further analysis.

Mycoplasma Laboratorium

Venter is seeking to patent the first life form created by humanity, possibly to be named *Mycoplasma laboratorium*. There is speculation that this line of research could lead to producing bacteria that have been engineered to perform specific reactions, e.g. produce fuels, make medicines, combat global warming, etc. In May 2010, a team of scientists led by Venter became the first to successfully create what was described as "synthetic life". This was done by synthesizing a very long DNA molecule containing an entire bacterium genome, and introducing this into another cell, analogous to the accomplishment of Eckard Wimmer's group, who synthesized and ligated an RNA virus genome and "booted" it in cell lysate. The single-celled organism contains four "watermarks" written into its DNA to identify it as synthetic and to help trace its descendants. The watermarks include

1. Code table for entire alphabet with punctuations
2. Names of 46 contributing scientists
3. Three quotations
4. The web address for the cell.

Genetic Distance

Genetic distance refers to the genetic divergence between species or between populations within a species. It considers a variety of parameters used to measure the genetic distance. Smaller genetic distances indicate a close genetic relationship whereas large genetic distances indicate a more distant genetic relationship. Genetic distance can be used to compare the genetic similarity between different species, such as humans and chimpanzees. Within a species genetic distance can be used to measure the divergence between different sub-species.

In its simplest form, the genetic distance between two populations is the difference in frequencies of a trait. For example the frequency of RH negative individuals is 50.4% among Basques, 41.2% in France and 41.1% in England. Thus the genetic difference between the Basques and French is 9.2% and the genetic difference between the French and the English is 0.1% for the RH negative trait. The genetic distance of several individual traits can then be averaged to compute an overall genetic distance.

Measures of Genetic Distance

There are several measures used to indicate genetic distance. These include:

Fixation Index

A commonly used measure of genetic distance is the fixation index which varies between 0 and 1. A value of 0 indicates that two populations are genetically identical whereas a value of 1 indicates that two populations are different species.

Nei's Standard Genetic Distance

This measure assumes that genetic differences arise due to mutations and genetic drift.

Cavalli-Sforza and Edwards 1967

This measure assumes that genetic differences arise due to genetic drift only.

Reynolds, Weir, and Cockerham's 1983

This measure assumes that genetic differences arise due to genetic drift only.

Noncoding DNA

In genetics, noncoding DNA describes components of an organism's DNA sequences that do not encode for protein sequences. In many eukaryotes, a large percentage of an organism's total genome size is noncoding DNA, although the amount of noncoding DNA, and the proportion of coding versus noncoding DNA varies greatly between species. Much of this DNA has no known biological function and at one time was sometimes referred to as "Junk DNA".

However, many types of noncoding DNA sequences do have known biological functions, including the transcriptional and translational regulation of protein-coding sequences. Other noncoding sequences have likely but as-yet undetermined function, an inference from high levels of homology and conservation seen in sequences that do not encode proteins but appear to be under heavy selective pressure.

Fraction of Noncoding Genomic DNA

The amount of total genomic DNA varies widely between organisms, and the proportion of coding and noncoding DNA within these genomes varies greatly as well. More than 98% of the human genome does not encode protein sequences, including most sequences within introns and most intergenic DNA. While overall genome size, and by extension the amount of noncoding DNA, are correlated to organism complexity, there are many exceptions.

For example, the genome of the unicellular *Polychaos dubium* (formerly known as *Amoeba dubia*) has been reported to contain more than 200 times the amount of DNA in humans. The pufferfish *Takifugu rubripes* genome is only about one eighth the size of the human genome, yet seems to have a comparable number of genes; approximately 90% of the *Takifugu* genome is noncoding DNA and most of the genome size difference appears to lie in the noncoding DNA.

The extensive variation in nuclear genome size among eukaryotic species is known as the C-value enigma or C-value paradox. About 80% of the nucleotide bases in the human genome may be transcribed, but transcription does not necessarily imply function.

Types of Noncoding DNA Sequences

Noncoding Functional RNA: Noncoding RNAs are functional RNA molecules that are not translated into protein. Examples of noncoding RNA include ribosomal RNA, transfer RNA, Piwi-interacting RNA and microRNA. MicroRNAs are predicted to control the translational activity of approximately 30% of all protein-coding genes in mammals and may be vital components in the progression or treatment of various diseases including cancer, cardiovascular disease, and the immune system response to infection.

Cis-regulatory Elements

Cis-regulatory elements are sequences that control the transcription of a gene. Cis-elements may be located in 5' or 3' untranslated regions or within introns. Promoters facilitate the transcription of a particular gene and are typically upstream of the coding region. Enhancer sequences may exert very distant effects on the transcription levels of genes.

Introns

Introns are non-coding sections of a gene, transcribed into the precursor mRNA sequence, but ultimately removed by RNA splicing during the processing to mature messenger RNA. Many introns appear to be mobile genetic elements. Studies of group I introns from *Tetrahymena* indicate that some introns appear to be selfish genetic elements, neutral to the host because they remove themselves from flanking exons during RNA processing and do not produce an expression bias between alleles with and without the intron.

Some introns do appear to have significant biological function, possibly through ribozyme functionality that may regulate tRNA and rRNA activity as well as protein-coding gene expression, evident in hosts that have become dependent on such introns over long periods of time; for example, the *trnL-intron* is found in all green plants and appears to have been vertically inherited for several billions of years, including more than a billion years within chloroplasts and an additional 2–3 billion years prior in the cyanobacterial ancestors of chloroplasts.

Pseudogenes

Pseudogenes are DNA sequences, related to known genes, that have lost their protein-coding ability or are otherwise no longer expressed in the cell. Pseudogenes arise from retrotransposition or genomic duplication of functional genes, and become "genomic fossils" that are nonfuctional due to mutations that prevent the transcription of the gene, such as within the gene promoter region, or fatally alter the translation of the gene, such as premature stop codons or frameshifts. Pseudogenes resulting from the retrotransposition of an RNA intermediate are known as processed pseudogenes; pseudogenes that arise from the genomic remains of duplicated genes or residues of inactivated are nonprocessed pseudogenes. While Dollo's Law suggests that the loss of function in pseudogenes is likely permanent, silenced genes may actually retain function for several million years and can be "reactivated" into protein-coding sequences and a substantial number of pseudogenes are actively transcribed. Because pseudogenes are presumed to evolve without evolutionary constraint, they can serve as a useful model of the type and frequencies various spontaneous genetic mutations.

Repeat Sequences, Transposons and Viral Elements

Transposons and retrotransposons are mobile genetic elements. Retrotransposon repeated sequences, which include long interspersed nuclear elements (LINEs) and short interspersed nuclear elements (SINEs), account for a large proportion of the genomic sequences in many species. Alu sequences, classified as a short interspersed nuclear element, are the most abundant mobile elements in the human genome. Some examples have been found of SINEs exerting transcriptional control of some protein-encoding genes. Endogenous retrovirus sequences are the product of reverse transcription of retrovirus genomes into the genomes of germ cells. Mutation within these retro-transcribed sequences can inactivate the viral genome.

Over 8% of the human genome is made up of (mostly decayed) endogenous retrovirus sequences, as part of the over 42% fraction that is recognizably derived of retrotransposons, while another 3% can be identified to be the remains of DNA transposons. Much of the remaining half of the genome that is currently without an explained origin is expected to have found its origin in transposable elements that were active so long ago (> 200 million years) that random mutations have rendered them unrecognizable. Genome size variation in at least two kinds of plants is mostly the result of retrotransposon sequences.

Telomeres

Telomeres are regions of repetitive DNA at the end of a chromosome, which provide protection from chromosomal deterioration during DNA replication.

Functions of Noncoding DNA

Many noncoding DNA sequences have very important biological functions. Comparative genomics reveals that some regions of noncoding DNA are highly conserved, sometimes on time-scales representing hundreds of millions of years, implying that these noncoding regions are under strong evolutionary pressure and positive selection. For example, in the genomes of humans and mice, which diverged from a common ancestor 65–75 million years ago, protein-coding DNA sequences account for only about 20% of conserved DNA, with the remaining majority of conserved DNA represented in noncoding regions. Linkage mapping often identifies chromosomal regions associated with a disease with no evidence of functional coding variants of genes within the region, suggesting that disease-causing genetic variants lie in the noncoding DNA. Some noncoding DNA sequences are genetic "switches" that do not encode proteins, but do regulate when and where genes are expressed. According to a comparative study of over 300 prokaryotic and over 30 eukaryotic genomes, eukaryotes appear to require a minimum amount of non-coding DNA. This minimum amount can be predicted using a growth model for regulatory genetic networks, implying that it is required for regulatory purposes. In humans the predicted minimum is about 5% of the total genome.

Some noncoding DNA sequences determine the expression levels of various genes. Other sequences of noncoding DNA determine where transcription factors attach. Some specific sequences of noncoding DNA may be features essential to chromosome structure, centromere function and homolog recognition in meiosis.

Noncoding DNA and Evolution

Shared sequences of apparently non-functional DNA are a major line of evidence for common descent. Pseudogene sequences appear to accumulate mutations more rapidly than coding sequences due to a loss of selective pressure. This allows for the creation of mutant alleles that incorporate new functions that may be favoured by natural selection; thus, pseudogenes can serve as raw material for evolution and can be considered "protogenes".

Junk DNA

Junk DNA, a term that was introduced in 1972 by Susumu Ohno, was a provisional label for the portions of a genome sequence for which no discernible function had been identified. According to a 1980 review in *Nature* by Leslie Orgel and Francis Crick, junk DNA has "little specificity and conveys little or no selective advantage to the organism".

The term is currently, however, an outdated concept, being used mainly in popular science and in a colloquial way in scientific publications, and may have slowed research into the biological functions of noncoding DNA. Several lines of evidence indicate that many "junk DNA" sequences are likely to have unidentified functional activity, and other sequences may have had functions in the past.

Still, a significant amount of the sequence of the genomes of eukaryotic organisms currently appears to fall under no existing classification other than "junk". For example, one experiment removed 0.1% of the mouse genome with no detectable effect on the phenotype. This result suggests that the removed DNA was largely nonfunctional. In addition, these sequences are enriched for the heterochromatic histone modification H3K9me3.

Phylogenetic Footprinting

Phylogenetic footprinting is a technique used to identify transcription factor binding sites (TFBS) within a non-coding region of DNA of interest by comparing it to the orthologous sequence in different species. When this technique is used with a large number of closely related species, this is called phylogenetic shadowing.

Researchers have found that non-coding pieces of DNA contain binding sites for regulatory proteins that govern the spatiotemporal expression of genes. These transcription factor binding sites (TFBS), or regulatory motifs, have proven hard to identify, primarily because they are short in length, and can show sequence variation.

The importance of understanding transcriptional regulation to many fields of biology has led researchers to develop strategies for predicting the presence of TFBS, many of which have led to publicly available databases. One such technique is Phylogenetic Footprinting.

Phylogenetic footprinting relies upon two major concepts:

1. The function and DNA binding preferences of transcription factors are well-conserved between diverse species.
2. Important non-coding DNA sequences that are essential for regulating gene expression will show differential selective pressure. A slower rate of change occurs in TFBS than in other, less critical, parts of the non-coding genome.

History

Phylogenetic footprinting was first used and published by Tagle et al. in 1988, which allowed researchers to predict evolutionary conserved cis-regulatory elements responsible for embryonic ε and γ globulin gene expression in primates. Before phylogenetic footprinting, DNase footprinting was used, where protein would be bound to DNA transcription factor binding sites (TFBS) protecting it from DNase digestion.

One of the problems with this technique was the amount of time and labour it would take. Unlike DNase footprinting, phylogenetic footprinting relies on evolutionary constraints within the genome, with the "important" parts of the sequence being conserved among the different species.

Protocol

It is important when using this technique to decide which genome your sequence should be aligned to. More divergent species will have less sequence homology between orthologous genes. Therefore, the key is to pick species that are related enough to detect homology, but divergent enough to maximize non-alignment "noise". Step wise approach to Phylogenetic footprinting consists of :

1. One should decide on the gene of interest.
2. Carefully choose species with orthologous genes.
3. Decide on the length of the upstream or maybe downstream region to be looked at.
4. Align the sequences.
5. Look for conserved regions and analyse them.

Not all TFBS are Found

Not all transcription binding sites can be found using phylogenetic footprinting due to the statistical nature this technique. Here are several reasons why some TFBS are not found:

Species Specific Binding Sites

Some binding sites seem to have no significant matches in most other species. Therefore, detecting these sites by phylogenetic footprinting is likely impossible unless a large number of closely related species are available.

Very Short Binding Sites

Some binding sites show excellent conservation, but just in a shorter region than the ones we looked for. Such short motifs (e.g., GC-box) are often happened by chance in nonfunctional sequences and detecting these motifs would be challenging.

Less Specific Binding Factors

Some binding sites show some conservation but have had insertions or deletions. It is not obvious if these sequences with insertions or deletions are still functional.

Though they may still be functional if the finding factor is less specific (or less 'picky' if you will). Because deletions and insertions are rare in binding sites, considering insertions and deletions in the sequence would detect a few more true TFBSs, but it could likely include many more false positives.

Not Enough Data

Some motifs are quite well conserved, but they are statistically insignificant a specific dataset. The motif might have appeared in different species by chance. These motifs could be detected if sequences from more organisms are available. So this will be less of a problem in the future.

Compound Binding Regions

Some transcription factors bind as dimers. Therefore, their binding sites may consist of two conserved regions, separated by a few variable nucleotides. Because of the variable internal sequence, the motif cannot be detected. However, if we could use a program to search for motifs containing a variable sequence in the middle, without counting mutations, these motifs could be discovered.

Accuracy

It is important to keep in mind that not all conserved sequences are under selection pressure. To eliminate false positives statistical analysis must be performed that will show that the motifs reported have a mutation rate meaningfully less than that of the surrounding nonfunctional sequence. Moreover, results could be more accurate if the prior knowledge about the sequence is considered. For example, some regulatory elements are repeated 15 times in a promoter region (e.g., some metallothionein promoters have up to 15 metal response elements (MREs)). Thus, to eliminate false motifs with inconsistent order across species, the orientation and order of regulatory elements in a promoter region should be the same in all species. This type of information could help us to identify regulatory elements that are not adequately conserved but occur in several copies in the input sequence.

Transcriptome

The transcriptome is the set of all RNA molecules, including mRNA, rRNA, tRNA, and other non-coding RNA produced in one or a population of cells. The term can be applied to the total set of transcripts in a given organism, or to the specific subset of transcripts present in a particular cell type. Unlike the genome, which is roughly fixed for a given cell line (excluding mutations), the transcriptome can vary with external environmental conditions. Because it includes all *mRNA* transcripts in the cell, the transcriptome reflects the genes that are being actively expressed at any given time, with the exception of mRNA degradation phenomena such as transcriptional attenuation. The study of *transcriptomics*, also referred to as expression profiling, examines the expression level of mRNAs in a given cell population, often using high-throughput techniques based on DNA microarray technology. The use of next-generation sequencing technology to study the transcriptome at the nucleotide level is known as RNA-Seq. Transcriptomics is the branch of molecular biology that deals with the study of messenger RNA molecules produced in an individual or population of a particular cell type.

Applications and Analysis

The transcriptomes of stem cells and cancer cells are of particular interest to researchers who seek to understand the processes of cellular differentiation and carcinogenesis. A number of organism-specific transcriptome databases have been constructed and annotated to aid in the identification of genes that are differentially expressed in distinct cell populations or subtypes; however, the analysis of relative mRNA expression levels can be complicated by the fact that relatively small changes in mRNA expression can produce large changes in the total amount of the corresponding protein present in the cell. One analysis method, known as Gene Set Enrichment Analysis, identifies coregulated gene networks rather than individual genes that are up-or down-regulated in different cell populations.

mRNA Regulation

Although microarray studies can reveal the relative amounts of different mRNAs in the cell, levels of mRNA are not directly proportional to the expression level of the proteins they code for. The number of protein molecules synthesized using a given mRNA molecule as a template is highly dependent on translation-initiation features of the mRNA sequence; in particular, the ability of the translation initiation sequence is a key determinant in the recruiting of ribosomes for protein translation. The complete protein complement of a cell or organism is known as the proteome.

A study of 158,807 mouse transcripts revealed that 4520 of these transcripts form antisense partners that are base pair complementary to the exons of genes. These results raise the possibility that significant numbers of "antisense RNA-coding genes" might participate in the regulation of the levels of expression of protein-coding mRNAs.

The Genographic Project

The Genographic Project, launched on April 13, 2005 by the National Geographic Society and IBM, is a multi-year genetic anthropology study that aims to map historical human migration patterns by collecting and analysing DNA samples from hundreds of thousands of people from around the world.

Overview

Field researchers at 11 regional centres around the world collect DNA samples from indigenous populations. The project also sells self-testing kits: for US$100 anyone in the world can order a kit with which a mouth scraping (buccal swab) is obtained, analysed and the DNA information placed on an Internet accessible database. The genetic markers on mitochondrial DNA (HVR1) and Y-chromosomes (12 microsatellite markers and haplogroup-defining SNPs) are used to trace the participant's distant ancestry, and each customer is provided with their genetic history. As of April 2010 more than 350,000 people had bought a test kit, and the success of the project has spawned a broader interest in direct-to-consumer genetic testing.

The Genographic Project is undertaking widespread consultation with indigenous groups from around the world. Genographic Project public participation kits are processed by Family Tree DNA (FTDNA) using the Arizona Research Labs at the University of Arizona. The project is a privately-funded, not-for-profit collaboration between the National Geographic Society, IBM and the Waitt Family Foundation. Part of the proceeds from the sale of self-testing kits support the Genographic Project's ongoing DNA collection, but the majority are ploughed into a Legacy Fund to be spent on cultural preservation projects nominated by indigenous communities.

Team Members

Team members include:

- Spencer Wells, project director (National Geographic Explorer-in-Residence)
- Jin Li, principal investigator, East Asia
- Theodore Schurr, principal investigator, North America
- Fabricio Santos, principal investigator, South America
- Jaume Bertranpetit, David Comas and Lluis Quintana-Murci, principal investigators, Western Europe and Central Europe
- Pierre Zalloua, principal investigator, Middle East and Northern Africa
- Himla Soodyall, principal investigator, Sub-Saharan Africa
- Elena Balanovska, principal investigator, North Eurasia
- Ramasamy Pitchappan, principal investigator, India
- Alan Cooper, principal investigator, Ancient DNA
- John Mitchell, principal investigator, Australia and New Zealand
- Lisa Matisoo-Smith, principal investigator, Oceania
- Ajay Royyuru, head of computational biology, IBM
- Simon Longstaff, advisory board chair (director of the St James Ethics Centre)
- Meave Leakey, advisory board member
- Merritt Ruhlen, advisory board member
- Colin Renfrew, advisory board member
- Luigi Luca Cavalli-Sforza, advisory board member
- Wade Davis, advisory board member.

Use of Genetic Markers

The Genographic Project relies on the identification of genetic markers. Most human DNA is a shuffled combination of genetic material passed down the generations. There are, however, parts of the human genome that pass unshuffled from parent to child. These segments of DNA are only changed by occasional mutations—random spelling mistakes in the genetic code. When these spelling mistakes are passed down to succeeding generations, they become markers of descent.

Different populations have different genetic markers, and by following them through the generations scientists are able to identify the different branches of the human tree, all the way back to their common African root.

Indigenous populations provide geographical and cultural context to the genetic markers in their DNA. These clues can help recreate past migration patterns.

Criticism

Shortly after the announcement of the project in April 2005, the Indigenous Peoples Council on Biocolonialism, (IPCB), released a statement protesting about the project, its connections with the HGDP, and called for a boycott of IBM, Gateway Computers, and National Geographic. Around May 2006, the United Nations Permanent Forum on Indigenous Issues (UNPFII) recommended suspending the project. Concerns were that the knowledge gleaned from the research could clash with long held beliefs leading to the destruction of their culture. They also feared that it could endanger land rights and other benefits.

In May 2006, the representatives of Indigenous went to UNPFII contesting any involvement in the testing. "The Genographic Project is exploitative and unethical because it will use Indigenous peoples as subjects of scientific curiosity in research that provides no benefit to Indigenous peoples, yet subjects them to significant risks. Researchers will take blood or other bodily tissue samples for their own use in order to further their own speculative theories of human history." UNPFII conducted investigations into the objectives of the Genographic Project, and concluded that since the project was "conceived and has been initiated without appropriate consultation with or regard for the risks to its subjects, the Indigenous peoples, the Council for Responsible Genetics concludes that the Indigenous peoples' representatives are correct and that the Project should be immediately suspended.

As of December 2006 some federally recognized tribes in North America have declined to take part. "What the scientists are trying to prove is that we're the same as the Pilgrims except we came over several thousand years before," said Maurice Foxx, chairman of the Massachusetts Commission on Indian Affairs and a member of the Mashpee Wampanoag. "Why should we give them that openly?" However, more than 50,000 indigenous participants from the Americas, Africa, Asia, Europe and Oceania had joined the project as of April 2010. As of 2010, the consensus is that the Genographic Project's benefits of providing genetic knowledge to humanity have created stronger ties among various cultural groups. While indigenous rights groups' concerns have been addressed, other cultural interest groups likewise have asserted their rights. Various extended family organizations worldwide support the Genographic Project. They have found connections to their places of heritage. They have discovered kinship relations to other ethnic groups and nationalities, creating stronger and more peaceful ties with other polities. Participants have discovered their "roots," representing an inalienable right to cultural heritage.

4

Cell Cycle Regulation in Cloning

The cell cycle, or cell-division cycle, is the series of events that takes place in a cell leading to its division and duplication (replication). In cells without a nucleus (prokaryotic), the cell cycle occurs via a process termed binary fission. In cells with a nucleus (eukaryotes), the cell cycle can be divided in three periods: interphase—during which the cell grows, accumulating nutrients needed for mitosis and duplicating its DNA—and the mitosis (M) phase, during which the cell splits itself into two distinct cells, often called "daughter cells" and the final phase, Cytokinesis, where the new cell is completely divided.

The cell-division cycle is a vital process by which a single-celled fertilized egg develops into a mature organism, as well as the process by which hair, skin, blood cells, and some internal organs are renewed.

Phases

The cell cycle consists of four distinct phases: G_1 phase, S phase (synthesis), G_2 phase (collectively known as interphase) and M phase (mitosis). M phase is itself composed of two tightly coupled processes: mitosis, in which the cell's chromosomes are divided between the two daughter cells, and cytokinesis, in which the cell's cytoplasm divides in half forming distinct cells. Activation of each phase is dependent on the proper progression and completion of the previous one. Cells that have temporarily or reversibly stopped dividing are said to have entered a state of quiescence called G_0 phase.

After cell division, each of the daughter cells begin the interphase of a new cycle. Although the various stages of interphase are not usually morphologically distinguishable, each phase of the cell cycle has a distinct set of specialised biochemical processes that prepare the cell for initiation of cell division.

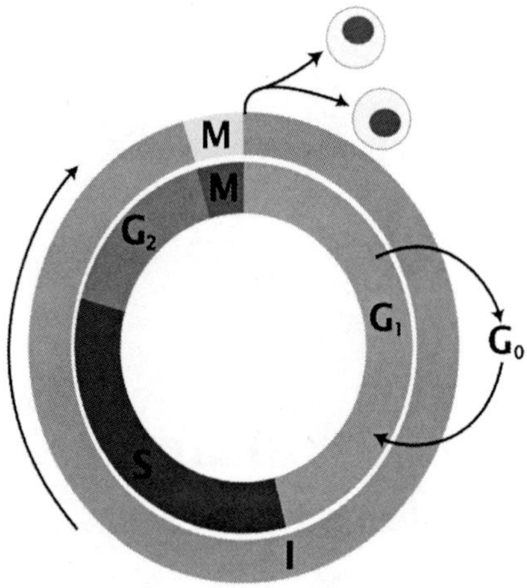

Figure : *Schematic of the cell cycle. outer ring: I = Interphase, M = Mitosis; inner ring: M = Mitosis, G1 = Gap 1, G2 = Gap 2, S = Synthesis; not in ring: G0 = Gap 0/Resting.*

Resting (G_0 Phase)

The term "post-mitotic" is sometimes used to refer to both quiescent and senescent cells. Nonproliferative cells in multicellular eukaryotes generally enter the quiescent G_0 state from G_1 and may remain quiescent for long periods of time, possibly indefinitely (as is often the case for neurons). This is very common for cells that are fully differentiated. Cellular senescence is a state that occurs in response to DNA damage or degradation that would make a cell's progeny nonviable; it is often a biochemical alternative to the self-destruction of such a damaged cell by apoptosis. As well, growth and division of such a cell could become cancerous.

Interphase

Before a cell can enter cell division, it needs to take in nutrients. All of the preparations are done during the interphase. Interphase proceeds in three stages, G1, S, and G2. Cell division operates in a cycle. Therefore, interphase is preceded by the previous cycle of mitosis and cytokinesis. Interphase is also known as preparatory phase, in this stage nucleus and cytochrome division does not occur.

G_1 Phase

The first phase within interphase, from the end of the previous M phase until the beginning of DNA synthesis is called G_1 (G indicating *gap*). It is also called the growth phase. During this phase the biosynthetic activities of the cell, which had been considerably slowed down during M phase, resume at a high rate. This phase is marked by synthesis of various enzymes that are required in S phase, mainly those needed for DNA replication. Duration of G_1 is highly variable, even among different cells of the same species. What also happens is that the chromosomes duplicate.

S Phase

The ensuing S phase starts when DNA synthesis commences; when it is complete, all of the chromosomes have been replicated, i.e., each chromosome has two (sister) chromatids. Thus, during this phase, the amount of DNA in the cell has effectively doubled, though the ploidy of the cell remains the same. Rates of RNA transcription and protein synthesis are very low during this phase. An exception to this is histone production, most of which appears during the S phase.

G_2 Phase

The cell then enters the G_2 phase, which lasts until the cell enters mitosis. Again, significant biosynthesis occurs during this phase, mainly involving the production of microtubules, which are required during the process of mitosis. Inhibition of protein synthesis during G_2 phase prevents the cell from undergoing mitosis.

Mitosis (M Phase/Mitotic Phase)

The relatively brief *M phase* consists of nuclear division (karyokinesis). The M phase has been broken down into several distinct phases, sequentially known as:

- prophase,
- metaphase,
- anaphase,
- telophase
- cytokinesis (strictly speaking, cytokinesis is not part of mitosis but is an event that directly follows mitosis in which cytoplasm is divided into two daughter cells).

Mitosis is the process by which a eukaryotic cell separates the chromosomes in its cell nucleus into two identical sets in two nuclei. It is generally followed immediately by cytokinesis, which divides the nuclei, cytoplasm, organelles and cell membrane into two cells containing roughly equal shares of these cellular components. Mitosis and cytokinesis together define the mitotic (M) phase of the cell cycle - the division of the mother cell into two daughter cells, genetically identical to each other and to their parent cell. This accounts for approximately 10% of the cell cycle. Mitosis occurs exclusively in eukaryotic cells, but occurs in different ways in different species. For example, animals undergo an "open" mitosis, where the nuclear envelope breaks down before the chromosomes separate, while fungi such as *Aspergillus nidulans* and *Saccharomyces cerevisiae* (yeast) undergo a "closed" mitosis, where chromosomes divide within an intact cell nucleus. Prokaryotic cells, which lack a nucleus, divide by a process called binary fission.

The process of mitosis is complex and highly regulated. The sequence of events is divided into phases, corresponding to the completion of one set of activities and the start of the next. These stages are prophase, prometaphase, metaphase, anaphase and telophase. During the process of mitosis the pairs of chromosomes condense and attach to fibres that pull the sister chromatids to opposite sides of the cell. The cell then divides in cytokinesis, to produce two identical daughter cells. Because cytokinesis usually occurs in conjunction with mitosis, "mitosis" is often used interchangeably with "M phase". However, there are many cells where mitosis and cytokinesis occur separately, forming single cells with multiple nuclei. This occurs most notably among the fungi and slime moulds, but is found in various different groups. Even in animals, cytokinesis and mitosis may occur independently, for instance during certain stages of fruit fly embryonic development. Errors in mitosis can either kill a cell through apoptosis or cause mutations that may lead to cancer.

Regulation of Eukaryotic cell Cycle

Regulation of the cell cycle involves processes crucial to the survival of a cell, including the detection and repair of genetic damage as well as the prevention of uncontrolled cell division. The molecular events that control the cell cycle are ordered and directional; that is, each process occurs in a sequential fashion and it is impossible to "reverse" the cycle.

Role of Cyclins and CDKs

Two key classes of regulatory molecules, cyclins and cyclin-dependent kinases (CDKs), determine a cell's progress through the cell cycle. Leland H. Hartwell, R. Timothy Hunt, and Paul M. Nurse won the 2001 Nobel Prize in Physiology or Medicine for their discovery of these central molecules. Many of the genes encoding cyclins and CDKs are conserved among all eukaryotes, but in general more complex organisms have more elaborate cell cycle control systems that incorporate more individual components. Many of the relevant genes were first identified by studying yeast, especially *Saccharomyces cerevisiae*; genetic nomenclature in yeast dubs many these genes *cdc* (for "cell division cycle") followed by an identifying number, e.g., *cdc25* or *cdc20*. Cyclins form the regulatory subunits and CDKs the catalytic subunits of an activated heterodimer; cyclins have no catalytic activity and CDKs are inactive in the absence of a partner cyclin. When activated by a bound cyclin, CDKs perform a common biochemical reaction called phosphorylation that activates or inactivates target proteins to orchestrate coordinated entry into the next phase of the cell cycle. Different cyclin-CDK combinations determine the downstream proteins targeted. CDKs are constitutively expressed in cells whereas cyclins are synthesised at specific stages of the cell cycle, in response to various molecular signals.

General Mechanism of Cyclin-CDK Interaction

Upon receiving a pro-mitotic extracellular signal, G_1 cyclin-CDK complexes become active to prepare the cell for S phase, promoting the expression of transcription factors that in turn promote the expression of S cyclins and of enzymes required for DNA replication. The G_1 cyclin-CDK complexes also promote the degradation of molecules that function as S phase inhibitors by targeting them for ubiquitination. Once a protein has been ubiquitinated, it is targeted for proteolytic degradation by the proteasome.

Active S cyclin-CDK complexes phosphorylate proteins that make up the pre-replication complexes assembled during G_1 phase on DNA replication origins. The phosphorylation serves two purposes: to activate each already-assembled pre-replication complex, and to prevent new complexes from forming. This ensures that every portion of the cell's genome will be replicated once and only once.

The reason for prevention of gaps in replication is fairly clear, because daughter cells that are missing all or part of crucial genes will die. However, for reasons related to gene copy number effects, possession of extra copies of certain genes is also deleterious to the daughter cells. Mitotic cyclin-CDK complexes, which are synthesized but inactivated during S and G_2 phases, promote the initiation of mitosis by stimulating downstream proteins involved in chromosome condensation and mitotic spindle assembly.

A critical complex activated during this process is a ubiquitin ligase known as the anaphase-promoting complex (APC), which promotes degradation of structural proteins associated with the chromosomal kinetochore. APC also targets the mitotic cyclins for degradation, ensuring that telophase and cytokinesis can proceed.

Specific Action of Cyclin-CDK Complexes

Cyclin D is the first cyclin produced in the cell cycle, in response to extracellular signals (e.g. growth factors). Cyclin D binds to existing CDK4, forming the active cyclin D-CDK4 complex. Cyclin D-CDK4 complex in turn phosphorylates the retinoblastoma susceptibility protein (Rb). The hyperphosphorylated Rb dissociates from the E2F/DP1/Rb complex (which was bound to the E2F responsive genes, effectively "blocking" them from transcription), activating E2F. Activation of E2F results in transcription of various genes like cyclin E, cyclin A, DNA polymerase, thymidine kinase, etc. Cyclin E thus produced binds to CDK2, forming the cyclin E-CDK2 complex, which pushes the cell from G_1 to S phase (G_1/S transition). Cyclin B along with cdc2 (cdc2 - fission yeasts (CDK1 - mammalia)) forms the cyclin B-cdc2 complex, which initiates the G_2/M transition. Cyclin B-cdc2 complex activation causes breakdown of nuclear envelope and initiation of prophase, and subsequently, its deactivation causes the cell to exit mitosis.

Inhibitors

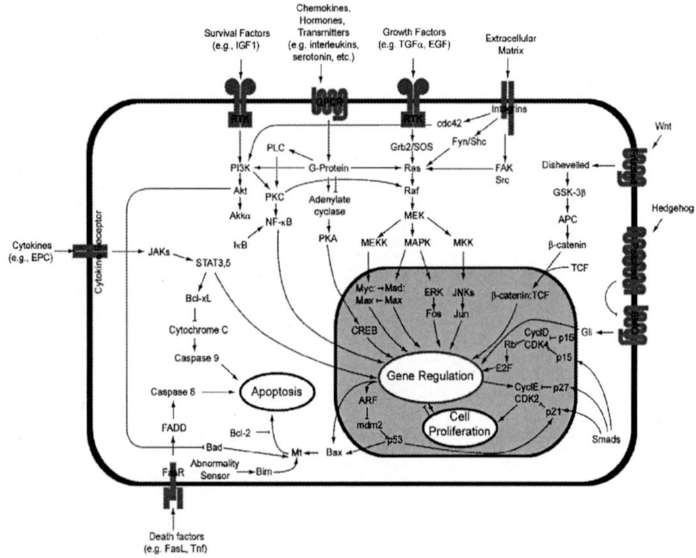

Figure : *Overview of signal transduction pathways involved in apoptosis, also known as "programmed cell death".*

Two families of genes, the *cip/kip* family (*CDK interacting protein/Kinase inhibitory protein*) and the INK4a/ARF (*In*hibitor of *K*inase 4/*A*lternative *R*eading *F*rame) prevent the progression of the cell cycle. Because these genes are instrumental in prevention of tumor formation, they are known as tumor suppressors. The *cip/kip* family includes the genes p21, p27 and p57. They halt cell cycle in G_1 phase, by binding to, and inactivating, cyclin-CDK complexes. p21 is activated by p53 (which, in turn, is triggered by DNA damage e.g. due to radiation). p27 is activated by Transforming Growth Factor â (TGF â), a growth inhibitor.

The INK4a/ARF family includes p16INK4a, which binds to CDK4 and arrests the cell cycle in G_1 phase, and p14arf which prevents p53 degradation. Synthetic inhibitors of Cdc25 could also be useful for the arrest of cell cycle and therefore be useful as antineoplastic and anticancer agents.

Transcriptional Regulatory Network

Evidence suggests that a semi-autonomous transcriptional network acts in concert with the CDK-cyclin machinery to regulate the cell cycle. Several gene expression studies in *Saccharomyces cerevisiae* have identified approximately 800 to 1200 genes that change expression over the course of the cell cycle; they are transcribed at high levels at specific points in the cell cycle, and remain at lower levels throughout the rest of the cell cycle. While the set of identified genes differs between studies due to the computational methods and criterion used to identify them, each study indicates that a large portion of yeast genes are temporally regulated.

Many periodically expressed genes are driven by transcription factors that are also periodically expressed. One screen of single-gene knockouts identified 48 transcription factors (about 20% of all non-essential transcription factors) that show cell cycle progression defects. Genome-wide studies using high throughput technologies have identified the transcription factors that bind to the promoters of yeast genes, and correlating these findings with temporal expression patterns have allowed the identification of transcription factors that drive phase-specific gene expression. The expression profiles of these transcription factors are driven by the transcription factors that peak in the prior phase, and computational models have shown that a CDK-autonomous network of these transcription factors is sufficient to produce steady-state oscillations in gene expression). Experimental evidence also suggests that gene expression can oscillate with the period seen in dividing wild-type cells independently of the CDK machinery. Orlando *et al.* used microarrays to measure the expression of a set of 1,271 genes that they identified as periodic in both wild type cells and cells lacking all S-phase and mitotic cyclins (*clb1,2,3,4,5,6*). Of the 1,271 genes assayed, 882 continued to be expressed in the cyclin-deficient cells at the same time as in the wild type cells, despite the fact that the cyclin-deficient cells arrest at the border between G1 and S phase. However, 833 of the genes assayed changed behaviour between the wild type and mutant cells, indicating that these genes are likely directly or indirectly regulated by the CDK-cyclin machinery. Some genes that continued to be expressed on time in the mutant cells were also expressed at different levels in the mutant and wild type cells. These findings suggest that while the transcriptional network may oscillate independently of the CDK-cyclin oscillator, they are coupled in a manner that requires both to ensure the proper timing of cell cycle events. Other work indicates that phosphorylation, a post-translational modification, of cell cycle transcription factors by Cdk1 may alter the localisation or activity of the transcription factors in order to tightly control timing of target genes (Ubersax *et al.* 2003; Sidorova *et al.* 1995; White *et al.* 2009).

While oscillatory transcription plays a key role in the progression of the yeast cell cycle, the CDK-cyclin machinery operates independently in the early embryonic cell cycle. Before the midblastula transition, zygotic transcription does not occur and all needed proteins, such as the B-type cyclins, are translated from maternally loaded mRNA.

Checkpoints

Cell cycle checkpoints are used by the cell to monitor and regulate the progress of the cell cycle. Checkpoints prevent cell cycle progression at specific points, allowing verification of necessary phase processes and repair of DNA damage. The cell cannot proceed to the next phase until checkpoint requirements have been met. Several checkpoints are designed to ensure that damaged or incomplete DNA is not passed on to daughter cells. Two main checkpoints exist: the G_1/S checkpoint and the G_2/M checkpoint. G_1/S transition is a rate-limiting step in the cell cycle and is also known as restriction point. An alternative model of the cell cycle response to DNA damage has also been proposed, known as the postreplication checkpoint. p53 plays an important role in triggering the control mechanisms at both G_1/S and G_2/M checkpoints.

Role in Tumor Formation

A disregulation of the cell cycle components may lead to tumor formation. As mentioned above, some genes like the cell cycle inhibitors, RB, p53 etc., when they mutate, may cause the cell to multiply uncontrollably, forming a tumor. Although the duration of cell cycle in tumor cells is equal to or longer than that of normal cell cycle, the proportion of cells that are in active cell division (versus quiescent cells in G_0 phase) in tumours is much higher than that in normal tissue. Thus there is a net increase in cell number as the number of cells that die by apoptosis or senescence remains the same. The cells which are actively undergoing cell cycle are targeted in cancer therapy as the DNA is relatively exposed during cell division and hence susceptible to damage by drugs or radiation. This fact is made use of in cancer treatment; by a process known as debulking, a significant mass of the tumor is removed which pushes a significant number of the remaining tumor cells from G_0 to G_1 phase (due to increased availability of nutrients, oxygen, growth factors etc.). Radiation or chemotherapy following the debulking procedure kills these cells which have newly entered the cell cycle.

The fastest cycling mammalian cells in culture, crypt cells in the intestinal epithelium, have a cycle time as short as 9 to 10 hours. Stem cells in resting mouse skin may have a cycle time of more than 200 hours. Most of this difference is due to the varying length of G_1, the most variable phase of the cycle. M and S do not vary much. In general, cells are most radiosensitive in late M and G_2 phases and most resistant in late S.

For cells with a longer cell cycle time and a significantly long G_1 phase, there is a second peak of resistance late in G_1 The pattern of resistance and sensitivity correlates with the level of sulfhydryl compounds in the cell. Sulfhydryls are natural radioprotectors and tend to be at their highest levels in S and at their lowest near mitosis.

Synchronisation of Cell Cultures

Several methods can be used to synchronise cell cultures by halting the cell cycle at a particular phase. For example, serum starvation and treatment with thymidine or aphidicolin halt the cell in the G_1 phase, mitotic shake-off, treatment with colchicine and treatment with nocodazole halt the cell in M phase and treatment with 5-fluorodeoxyuridine halts the cell in S phase.

In an attempt to better define the mode of action of TCGF we have investigated the effect of complete removal of TCGF on the cell cycle kinetics of a TCGF-dependent murine CTL cloned line (B6.1). Cells that had been cultured in the presence or absence of TCGF were stained with propidium iodide, a DNA binding dye, and the distribution of DNA content was analysed by flow cytometry. Results indicated that the proportion of cells entering the S phase of the cell cycle started to decrease 6 hr after removal of TCGF from the culture medium. This fraction continued to decrease as a function of time, and after 30 hr without TCGF 93% of the cells were in the G1 phase of the cell cycle in comparison to 38% in control cultures containing TCGF. Direct evidence that B6.1 cells completed one cell cycle after removal of TCGF before they accumulated in the G1 phase was obtained with a technique combining the use of BUdR, a thymidine analog, and the DNA binding dye Hoechst 33342. When TCGF was added again to cultures of B6.1 cells that were arrested in the G1 phase, the majority of cells entered the S phase in a synchronous fashion after a lag phase of 10 to 12 hr. The duration of this lag phase was independent of the concentration of TCGF used to restimulate quiescent B6.1 cells. However, the number of cells entering S phase 10 hr after re-addition of TCGF was concentration dependent. Additional experiments indicated that the cytolytic activity was not dependent on TCGF, since no decrease in the activity of B6.1 cells was noted after these cells had been deprived of TCGF for up to 30 hr.

Chromosome replication in the asymmetrically dividing bacteria Caulobacter crescentus is discontinuous with the new, motile swarmer cell undergoing an obligatory presynthetic gap period (G1 period) of 60 min before the initiation of DNA synthesis and stalk formation. To examine the regulation of the cell division cycle at the molecular level, we have cloned the DNA chain elongation gene dnaC from a genomic DNA library constructed in cosmid vector pLAFR1-7. To ensure that the cloned sequence corresponded to dnaC, we isolated the gene by genetic complementation of the temperature-sensitive allele dnaC303 on DNA fragment that contained a Tn5 insertion element tightly linked by transduction to dnaC. The size of the dnaC gene was estimated to be 1,500 bp or less based on the pattern of complementation by subcloned restriction and BAL 31 deletion fragments. Nuclease S1 assays were used to map the transcription start site and to determine the pattern of dnaC expression in the cell cycle.

Large amounts of the dnaC transcript began to accumulate only in the late G1 period of the swarmer cell and then peaked early during chromosome replication. We confirmed that the gene is periodically transcribed by monitoring the rate of beta-galactosidase synthesis directed by a dnaC promoter-lacZ fusion in a synchronous cell culture. dnaC is the first C. crescentus cell cycle gene whose regulation has been reported, and the discontinuous pattern of its expression suggests that the DNA synthetic period in these dimorphic bacteria is regulated in part by the stage-specific expression of DNA replication genes.

Interphase

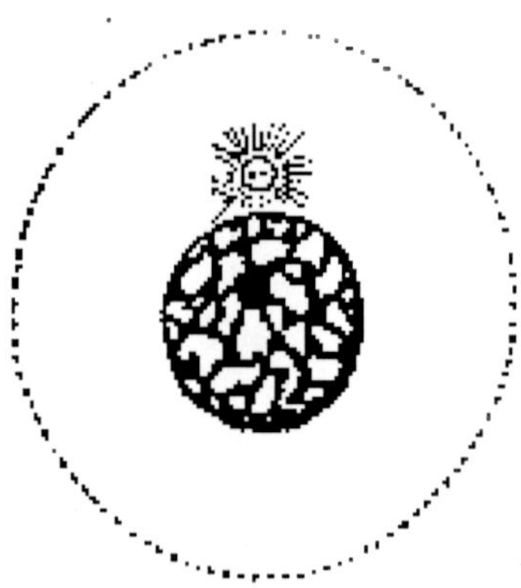

Figure : *An illustration of interphase in Gray's Anatomy. The chromatin has not yet condensed, and the cell is undergoing its normal functions.*

An image of the nucleus of a cell (HT1080) currently in interphase (probably G1). Note: Cytoplasm of this cell or the neighbouring cell is not visible, which is currently in the telophase of mitosis. Image taken using an optical microscope and DAPI staining of DNA.

Interphase is the phase of the cell cycle in which the cell spends the majority of its time and performs the majority of its purposes including preparation for cell division. In preparation for cell division, it increases its size and makes a copy of its DNA. Interphase is also considered to be the 'living' phase of the cell, in which the cell obtains nutrients, grows, reads its DNA, and conducts other "normal" cell functions. The majority of eukaryotic cells spend most of their time in interphase. Interphase does not describe a cell that is merely resting but is rather an active preparation for cell division. A common misconception is that interphase is the first stage of mitosis. However, since mitosis is the division of the nucleus, prophase is actually the first stage.

In interphase, the cell gets itself ready for mitosis or meiosis. Somatic cells, or normal diploid cells of the body, go through mitosis in order to reproduce themselves through cell division, whereas diploid germ cells (i.e., primary spermatocytes and primary oocytes) go through meiosis in order to create haploid gametes (i.e., sperm and ova) for the purpose of sexual reproduction.

Identification

Under a microscope, interphase can be recognised because the nuclear membrane is still intact, the chromatin has not yet condensed, and chromosomes are not visible, though the nucleolus may be visible as an enlarged dark spot. The centrioles and spindle fibres are also not yet visible, though the centrosome, which contains and organises them, may be visible near the nucleus.

Stages of Interphase

There are three stages of interphase, with each phase ending when a cellular checkpoint checks the accuracy of the stage's completion before proceeding to the next. The stages of interphase are:

- G_1 (Gap 1), in which the cell grows and functions normally. During this time, much protein synthesis occurs and the cell grows (to about double its original size) - more organelles are produced, increasing the volume of the cytoplasm. If the cell is not to divide again, it will remain in this phase.
- Synthesis (S), in which the cell duplicates its DNA (via semiconservative replication). This is also known as the Swanson phase.
- G_2 (Gap 2), in which the cell resumes its growth in preparation for mitosis.
- In addition, some cells that do not divide often or ever, enter a stage called G_0 (Gap zero), which is either a stage separate from interphase or an extended G_1 phase, which follows the restriction point, a cell cycle checkpoint found at the end of G_1.

The duration of time spent in interphase and in each stage of interphase is variable and depends on both the type of cell and the species of organism it belongs to. Most cells of adult mammals spend about 20 hours in interphase, this accounts for about 90% of the total time involved in cell division.

Interphase Within Sequences of Cellular Processes

Interphase and the cell Cycle

When G_2 is completed, the cell enters a relatively brief period of nuclear and cellular division, composed of mitosis and cytokinesis, respectively. After the successful completion of mitosis and cytokinesis, both resulting daughter cells re-enter G_1 of interphase.

In the cell cycle, interphase is proceeded by telophase and cytokinesis of the M phase. In alternative fashion, interphase is sometimes interrupted by G_0 phase, which, in some circumstances, may then end and be followed by the remaining stages of interphase. After the successful completion of the G_2 checkpoint, the final checkpoint in interphase, the cell proceeds to prophase, or in plants to preprophase, which is the first stage of mitosis. G_0 phase is viewed as either an extended G_1 phase where the cell is neither dividing nor preparing to divide and or as a distinct quiescent stage which occurs outside of the cell cycle.

Interphase and Other Cellular Processes

In gamete production interphase is succeeded by meiosis. In programmed cell death, interphase is followed or preempted by apoptosis.

Tissue Culture

Tissue culture is the growth of tissues and/or cells separate from the organism. This is typically facilitated via use of a liquid, semi-solid, or solid growth medium, such as broth or agar. Tissue culture commonly refers to the culture of animal cells and tissues, while the more specific term plant tissue culture is being named for the plants.

Historical Usage

In 1885 Wilhelm Roux removed a section of the medullary plate of an embryonic chicken and maintained it in a warm saline solution for several days, establishing the basic principle of tissue culture. In 1907 the zoologist Ross Granville Harrison demonstrated the growth of frog nerve cell processes in a medium of clotted lymph. In 1913, E. Steinhardt, C. Israeli, and R. A. Lambert grew vaccinia virus in fragments of guinea pig corneal tissue.

Modern Usage

In modern usage, tissue culture generally refers to the growth of cells from a tissue from a multicellular organism *in vitro*. These cells may be cells isolated from a donor organism, primary cells, or an immortalised cell line. The term tissue culture is often used interchangeably with cell culture. The literal meaning of tissue culture refers to the culturing of tissue pieces, i.e.

explant culture. Tissue culture is an important tool for the study of the biology of cells from multicellular organisms. It provides an *in vitro* model of the tissue in a well defined environment which can be easily manipulated and analysed.

Cell Culture

Cell culture is the complex process by which cells are grown under controlled conditions. In practice, the term "cell culture" has come to refer to the culturing of cells derived from singlecellular eukaryotes, especially animal cells. However, there are also cultures of plants, fungi and microbes, including viruses, bacteria and protists. The historical development and methods of cell culture are closely interrelated to those of tissue culture and organ culture.

Animal cell culture began when scientists removed a small amount of monkey kidney from a host organism and maintained it in a warm saline solution for several days, establishing the principle of tissue culture.

Concepts in Mammalian cell Culture

Isolation of Cells

Cells can be isolated from tissues for *ex vivo* culture in several ways. Cells can be easily purified from blood; however, only the white cells are capable of growth in culture. Mononuclear cells can be released from soft tissues by enzymatic digestion with enzymes such as collagenase, trypsin, or pronase, which break down the extracellular matrix. Alternatively, pieces of tissue can be placed in growth media, and the cells that grow out are available for culture. This method is known as explant culture. Cells that are cultured directly from a subject are known as primary cells. With the exception of some derived from tumours, most primary cell cultures have limited lifespan. After a certain number of population doublings (called the Hayflick limit), cells undergo the process of senescence and stop dividing, while generally retaining viability.

An established or immortalized cell line has acquired the ability to proliferate indefinitely either through random mutation or deliberate modification, such as artificial expression of the telomerase gene. Numerous cell lines are well established as representative of particular cell types.

Maintaining Cells in Culture

Cells are grown and maintained at an appropriate temperature and gas mixture (typically, 37°C, 5% CO_2 for mammalian cells) in a cell incubator. Culture conditions vary widely for each cell type, and variation of conditions for a particular cell type can result in different phenotypes being expressed. Aside from temperature and gas mixture, the most commonly varied factor in culture systems is the growth medium. Recipes for growth media can vary in pH, glucose concentration, growth factors, and the presence of other nutrients.

The growth factors used to supplement media are often derived from animal blood, such as calf serum. One complication of these blood-derived ingredients is the potential for contamination of the culture with viruses or prions, particularly in medical biotechnology applications. Current practice is to minimise or eliminate the use of these ingredients wherever possible and use chemically defined media, but this cannot always be accomplished. Alternative strategies involve sourcing the animal blood from countries with minimum BSE/TSE risk, such as Australia and New Zealand, and using purified nutrient concentrates derived from serum in place of whole animal serum for cell culture.

Plating density (number of cells per volume of culture medium) plays a critical role for some cell types. For example, a lower plating density makes granulosa cells exhibit estrogen production, while a higher plating density makes them appear as progesterone-producing theca lutein cells. Cells can be grown either in suspension or adherent cultures. Some cells naturally live in suspension, without being attached to a surface, such as cells that exist in the bloodstream. There are also cell lines that have been modified to be able to survive in suspension cultures so they can be grown to a higher density than adherent conditions would allow. Adherent cells require a surface, such as tissue culture plastic or microcarrier, which may be coated with extracellular matrix components to increase adhesion properties and provide other signals needed for growth and differentiation. Most cells derived from solid tissues are adherent. Another type of adherent culture is organotypic culture, which involves growing cells in a three-dimensional (3-D) environment as opposed to two-dimensional culture dishes. This 3D culture system is biochemically and physiologically more similar to *in vivo* tissue, but is technically challenging to maintain because of many factors (e.g. diffusion).

Cell Line Cross-contamination

Cell line cross-contamination can be a problem for scientists working with cultured cells. Studies suggest anywhere from 15–20% of the time, cells used in experiments have been misidentified or contaminated with another cell line. Problems with cell line cross-contamination have even been detected in lines from the NCI-60 panel, which are used routinely for drug-screening studies. Major cell line repositories, including the American Type Culture Collection (ATCC) and the German Collection of Microorganisms and Cell Cultures (DSMZ), have received cell line submissions from researchers that were misidentified by them. Such contamination poses a problem for the quality of research produced using cell culture lines, and the major repositories are now authenticating all cell line submissions. ATCC uses short tandem repeat (STR) DNA fingerprinting to authenticate its cell lines. To address this problem of cell line cross-contamination, researchers are encouraged to authenticate their cell lines at an early passage to establish the identity of the cell line. Authentication should be repeated before freezing cell line stocks, every two months during active culturing and before any publication of research data generated using the cell lines. Many methods are used to identify cell lines, including isoenzyme analysis, human lymphocyte antigen (HLA) typing, chromosomal analysis, karyotyping, morphology and STR analysis.

One significant cell-line cross contaminant is the immortal HeLa cell line.

Manipulation of Cultured Cells

As cells generally continue to divide in culture, they generally grow to fill the available area or volume. This can generate several issues:

- Nutrient depletion in the growth media
- Accumulation of apoptotic/necrotic (dead) cells
- Cell-to-cell contact can stimulate cell cycle arrest, causing cells to stop dividing, known as contact inhibition.
- Cell-to-cell contact can stimulate cellular differentiation.

Among the common manipulations carried out on culture cells are media changes, passaging cells, and transfecting cells. These are generally performed using tissue culture methods that rely on sterile technique. Sterile technique aims to avoid contamination with bacteria, yeast, or other cell lines. Manipulations are typically carried out in a biosafety hood or laminar flow cabinet to exclude contaminating micro-organisms. Antibiotics (e.g. penicillin and streptomycin) and antifungals (e.g.amphotericin B) can also be added to the growth media. As cells undergo metabolic processes, acid is produced and the pH decreases. Often, a pH indicator is added to the medium to measure nutrient depletion.

Media Changes

In the case of adherent cultures, the media can be removed directly by aspiration, and then is replaced.

Passaging Cells

Passaging (also known as subculture or splitting cells) involves transferring a small number of cells into a new vessel. Cells can be cultured for a longer time if they are split regularly, as it avoids the senescence associated with prolonged high cell density. Suspension cultures are easily passaged with a small amount of culture containing a few cells diluted in a larger volume of fresh media. For adherent cultures, cells first need to be detached; this is commonly done with a mixture of trypsin-EDTA; however, other enzyme mixes are now available for this purpose. A small number of detached cells can then be used to seed a new culture.

Transfection and Transduction

Another common method for manipulating cells involves the introduction of foreign DNA by transfection. This is often performed to cause cells to express a protein of interest. More recently, the transfection of RNAi constructs have been realised as a convenient mechanism for suppressing the expression of a particular gene/protein. DNA can also be inserted into cells using viruses, in methods referred to as transduction, infection or transformation. Viruses, as parasitic agents, are well suited to introducing DNA into cells, as this is a part of their normal course of reproduction.

Established Human Cell Lines

Cell lines that originate with humans have been somewhat controversial in bioethics, as they may outlive their parent organism and later be used in the discovery of lucrative medical treatments. In the pioneering decision in this area, the Supreme Court of California held in *Moore v. Regents of the University of California* that human patients have no property rights in cell lines derived from organs removed with their consent.

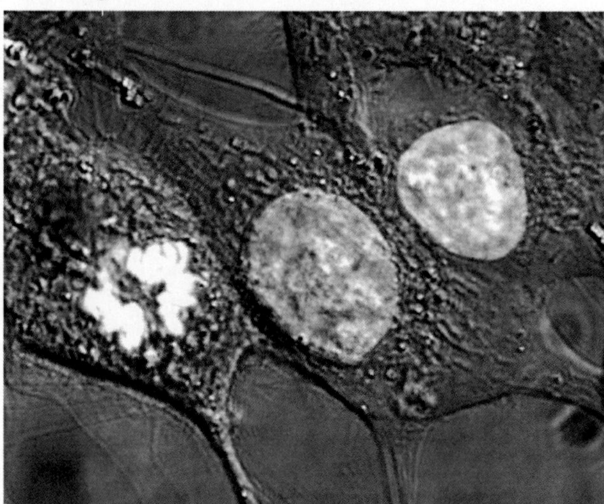

Figure : *Cultured HeLa cells have been stained with Hoechst turning their nuclei blue, and are one of the earliest human cell lines descended from Henrietta Lacks, who died of cervical cancer from which these cells originated.*

Generation of Hybridomas

It is possible to fuse normal cells with an immortalised cell line. This method is used to produce monoclonal antibodies. In brief, lymphocytes isolated from the spleen (or possibly blood) of an immunised animal are combined with an immortal myeloma cell line (B cell lineage) to produce a hybridoma which has the antibody specificity of the primary lymphoctye and the immortality of the myeloma. Selective growth medium (HA or HAT) is used to select against unfused myeloma cells; primary lymphoctyes die quickly in culture and only the fused cells survive. These are screened for production of the required antibody, generally in pools to start with and then after single cloning.

Applications of Cell Culture

Biological products produced by recombinant DNA (rDNA) technology in animal cell cultures include enzymes, synthetic hormones, immunobiologicals (monoclonal antibodies, interleukins, lymphokines), and anticancer agents. Although many simpler proteins can be produced using rDNA in bacterial cultures, more complex proteins that are glycosylated (carbohydrate-modified) currently must be made in animal cells. An important example of such a complex protein is the hormone erythropoietin.

The cost of growing mammalian cell cultures is high, so research is underway to produce such complex proteins in insect cells or in higher plants, use of single embryonic cell and somatic embryos as a source for direct gene transfer via particle bombardment, transit gene expression and confocal microscopy observation is one of its applications. It also offers to confirm single cell origin of somatic embryos and the asymmetry of the first cell division, which starts the process.

Cell Culture in two Dimensions

Research in tissue engineering, stem cells and molecular biology primarily involves cultures of cells on flat plastic dishes. This technique is known as two-dimensional (2D) cell culture, and was first developed by Wilhelm Roux who, in 1885, removed a portion of the medullary plate of an embryonic chicken and maintained it in warm saline for several days on a flat glass plate. With the advance of polymer technology, today's standard plastic dish for 2D cell culture, commonly known as the Petri dish, was invented by Julius Richard Petri, a German bacteriologist, who is generally credited with inventing the Petri dish while working as assistant to Robert Koch.

Tissue Culture and Engineering

Cell culture is a fundamental component of tissue culture and tissue engineering, as it establishes the basics of growing and maintaining cells *ex vivo*. The major application of human cell culture is in stem cell industry, where mesenchymal stem cells can be cultured and cryopreserved for future use.

Vaccines

Vaccines for polio, measles, mumps, rubella, and chickenpox are currently made in cell cultures. Due to the H5N1 pandemic threat, research into using cell culture for influenza vaccines is being funded by the United States government. Novel ideas in the field include recombinant DNA-based vaccines, such as one made using human adenovirus (a common cold virus) as a vector, and novel adjuvants.

Culture of Non-mammalian Cells

Plant Cell Culture Methods

Plant cell cultures are typically grown as cell suspension cultures in a liquid medium or as callus cultures on a solid medium. The culturing of undifferentiated plant cells and calli requires the proper balance of the plant growth hormones auxin and cytokinin.

Bacterial and Yeast Culture Methods

For bacteria and yeasts, small quantities of cells are usually grown on a solid support that contains nutrients embedded in it, usually a gel such as agar, while large-scale cultures are grown with the cells suspended in a nutrient broth.

Viral Culture Methods

The culture of viruses requires the culture of cells of mammalian, plant, fungal or bacterial origin as hosts for the growth and replication of the virus. Whole wild type viruses, recombinant viruses or viral products may be generated in cell types other than their natural hosts under the right conditions. Depending on the species of the virus, infection and viral replication may result in host cell lysis and formation of a viral plaque.

Common Cell Lines

Human cell lines

- HeLa
- National Cancer Institute's 60 cancer cell lines
- ESTDAB database
- DU145 (prostate cancer)
- Lncap (prostate cancer)
- MCF-7 (breast cancer)
- MDA-MB-438 (breast cancer)
- PC3 (prostate cancer)
- T47D (breast cancer)
- THP-1 (acute myeloid leukemia)
- U87 (glioblastoma)

- SHSY5Y Human neuroblastoma cells, cloned from a myeloma
- Saos-2 cells (bone cancer).

Primate cell lines:

- Vero (African green monkey *Chlorocebus* kidney epithelial cell line initiated in 1962)

Rat tumor cell lines:

- GH3 (pituitary tumor)
- PC12 (pheochromocytoma)

Mouse cell lines:

- MC3T3 (embryonic calvarium)

Plant cell lines:

- Tobacco BY-2 cells (kept as cell suspension culture, they are model system of plant cell)

Other species cell lines:

- Zebrafish ZF4 and AB9 cells
- Madin-Darby canine kidney (MDCK) epithelial cell line
- Xenopus A6 kidney epithelial cells.

Biological Immortality

Biological immortality refers to a stable rate of mortality as a function of chronological age. Some individual cells and entire organisms in some species achieve this state either throughout their existence or after living long enough. This requires that death occur from injury or disease rather than deterioration, i.e., the absence of cellular senescence. However, this definition of immortality has been challenged in the new *Handbook of the Biology of Aging*, because the increase in rate of mortality as a function of chronological age may be negligible at extremely old ages, also referred to as the *late-life mortality plateau*.

The rate of mortality may cease to increase in old age, but in most cases that rate is typically very high. As a hypothetical example, there is only a 50% chance of a human surviving another year at age 110 or greater. The fact that some multicellular organisms do not experience senescence implies that aging is not an Aristotelian inevitability and refutes the prior hypothesis that aging follows a Gompertz model. No actual organism or individual cell is inviolably immortal (i.e. "invincible" or "indestructible"). Biologically immortal living things die, for example, upon receiving sufficient injury or otherwise having its body destroyed or diseased.

Cell Lines

Biologists have chosen the word immortal to designate cells that are not limited by the Hayflick limit (where cells no longer divide because of DNA damage or shortened telomeres). Prior to Leonard Hayflick, Alexis Carrel hypothesized that all normal somatic cells are immortal.

The term immortalization was first applied to cancer cells that expressed the telomere-lengthening enzyme telomerase, and thereby avoided apoptosis (cell death caused by intracellular mechanisms). Among the most commonly used cell lines are HeLa and Jurkat, both of which are immortalized cancer cell lines. Normal stem cells and germ cells can also be said to be immortal (when humans refer to the cell line). Immortal cell lines of cancer cells can be created by induction of oncogenes or loss of tumor suppressor genes. One way to induce immortality is through viral-mediated induction of the large T-antigen, commonly introduced through simian virus 40 (SV-40).

Organisms

Immortality may not be desirable for multicellular organisms, as the main controls over cancer are the apoptotic mechanisms.

Tardigrades

Tardigrades, otherwise known as "water bears", are highly resilient microscopic animals. Capable of surviving extremes such as heat, radiation, drought, and even the vacuum of space by going into suspended animation, where their metabolism slows to near zero and they simply wait out the harsh conditions until the environment is more favourable. The Russian Space Administration is planning a mission to launch water bears at Phobos to study the scientific theory of panspermia.

Bacteria

Bacteria are said to be biologically immortal, but only at the level of the colony. An individual bacterium can easily die. The two daughter bacteria resulting from cell division of a parent bacterium can be regarded as unique individuals or as members of a biologically "immortal" colony. The two daughter cells can be regarded as "rejuvenated" copies of the parent cell because damaged macromolecules have been split between the two cells and diluted. In the same way stem cells and gametes can be regarded as "immortal".

Hydra

Hydras are a genus of simple, fresh-water animals possessing radial symmetry and no post-mitotic cells. The fact that all cells continually divide allows defects and toxins to be "diluted". It has been suggested that hydras do not undergo senescence (aging), and as such are biologically immortal.

However, this does not explain how hydras are consequently able to maintain telomere lengths.

Jellyfish

Turritopsis nutricula is a small (5 millimeters (0.20 in)) species of jellyfish which uses transdifferentiation to replenish cells after sexual reproduction. This cycle can repeat indefinitely, potentially rendering it biologically immortal. It originated from the Caribbean sea, but has now spread around the world.

Lobsters

Older lobsters are more fertile than younger lobsters. Some scientists have claimed that they could effectively live indefinitely, barring injury, disease, capture, etc.

Life Extensionists

Some life extensionists, such as those who practice cryonics, have the hope that humans may someday become biologically immortal. Biogerontologist Marios Kyriazis suggested that biological immortality in humans is an inevitable consequence of natural evolution. . His ELPIs (Extreme Lifespans through Perpetual-equalising Interventions) theory proposes that the ability to attain indefinite lifespans is inherent in human biology, and that there will come a time when humans will continue to develop their intelligence and sophistication by living indefinitely, rather than through Darwinian evolution.

Cell Culture Assays

In Biomaterials Testing, a cell culture assay is any method which is used to assess the cytotoxicity of a material. This refers to the *in vitro* assessment of material to determine whether it releases toxic chemicals in sufficient quantities to kill cells either directly or indirectly through the inhibition of cell metabolic pathways. Cell culture evaluations are the precursor to whole animal studies and are a way to determine if significant cytotoxicity exists for the given material. Cell culture assays are standardised by ASTM, ISO, and BSI (British Standards Institution.)

Methods

Direct Contact Method

1. A near confluent layer of fibroblasts are prepared in a culture plate
2. Old cell culture media (agar generally) is removed
3. Fresh media is added
4. Material being tested is placed onto the cultures, which are incubated for 24 hours at 37 degrees Celsius

5. The material is removed
6. The culture media is removed
7. The remaining cells are fixed and stained, dead cells are lost during fixation and only the live cells are stained
8. The toxicity of the material is indicated by the absence of stained cells around the material.

Agar Diffusion Method

1. A near confluent layer of fibroblasts are prepared in a culture plate
2. Old cell culture media is removed
3. The cells are covered with a solution of 2% agar, which often contains red vital stain
4. When the agar solidifies the cells will have dispersed throughout its volume
5. The material is then placed on the surface of the agar and incubated for 24 hours at 37 degrees Celsius
6. Live cells take up the vital stain and retain it, dead cells do not
7. The toxicity of the material is evaluated by the loss of vital stain under and around the material
8. Surface microscopy is also needed to evaluate the material-cell interface.

Elution Method

1. A near confluent layer of fibroblasts are prepared in a culture plate
2. An extract of the material which is being tested is prepared using physiological saline or serum free media (the latter is generally preferred)
3. Extraction conditions are used which are appropriate for the type of exposure which the cells would receive in the *in vivo* environment if the material were to be implanted
4. The extract is placed on the cells and incubated for 48 hours at 37 degrees Celsius
5. After 48 hours the toxicity is evaluated using either a histochemical or vital stain.

Each method has its own advantages and disadvantages, and some are more suitable for certain applications than others. For example the direct contact method offers conditions which are most similar to the physiological environment but the cells are susceptible to trauma if the material moves. The agar diffusion method is good for materials with high densities and offers an even concentration gradient for potential toxicants, but there is a serious risk of the cells going into thermal shock when they are overlayed with agar. The elution method is best for applications which might require extra incubation time, but additional time and steps are required for preparing such a test.

In vitro biomaterials testing yields fundamental information about the behaviour of materials in contact with living cells, but cannot qualify or even accurately predict the performance of a material *in vivo*.

Electric Cell-substrate Impedance Sensing

Electric cell-substrate impedance sensing or ECIS (a trademark of Applied BioPhysics Inc.) refers to a non-invasive biophysical approach to monitor living animal cells *in vitro*, i.e. within a well-defined laboratory environment. In ECIS the cells are grown on the surface of small and planar gold-film electrodes, which are deposited on the bottom of a cell culture dish (petri dish). The AC impedance of the cell-covered electrode is then measured at one or several frequencies as a function of time. Due to the insulating properties of their membranes the cells behave like dielectric particles so that the impedance increases with increasing coverage of the electrode until a confluent (i.e. continuous) layer of cells is established. In confluent cell layers the measured impedance is mainly determined by the three-dimensional shape of the cells. If cell shape changes occur, the current pathways through and around the cell bodies change as well, leading to a corresponding increase or decrease of impedance. Thus, by recording time-resolved impedance measurements, cell shape changes can be followed in real time with sub-microscopic resolution and can be used for bioanalytic purposes.

As the shape of animal cells responds very sensitively to alterations in metabolism as well as chemical, biological or physical stimuli, the ECIS technique is applied in various different experimental settings in cell biological research laboratories. It can be used as a sensor in cytotoxicity studies, drug development or as a non-invasive means to follow cell adhesion to *in vitro* surfaces. Equipments based on the ECIS technique are also dedicated to monitor the chemokinetic activity of adherent cells spread on the electrode surface (micromotion) as well as their chemotactic activities in ECIS-based wound healing assays.

Plant Tissue Culture

Plant tissue culture is a collection of techniques used to maintain or grow plant cells, tissues or organs under sterile conditions on a nutrient culture medium of known composition. Plant tissue culture is widely used to produce clones of a plant in a method known as micropropagation. Different techniques in plant tissue culture may offer certain advantages over traditional methods of propagation, including:

- The production of exact copies of plants that produce particularly good flowers, fruits, or have other desirable traits.
- To quickly produce mature plants.
- The production of multiples of plants in the absence of seeds or necessary pollinators to produce seeds.
- The regeneration of whole plants from plant cells that have been genetically modified.
- The production of plants in sterile containers that allows them to be moved with greatly reduced chances of transmitting diseases, pests, and pathogens.
- The production of plants from seeds that otherwise have very low chances of germinating and growing, i.e.: orchids and nepenthes.
- To clean particular plants of viral and other infections and to quickly multiply these plants as 'cleaned stock' for horticulture and agriculture.

Plant tissue culture relies on the fact that many plant cells have the ability to regenerate a whole plant (totipotency). Single cells, plant cells without cell walls (protoplasts), pieces of leaves, or (less commonly) roots can often be used to generate a new plant on culture media given the required nutrients and plant hormones.

Techniques

Modern plant tissue culture is performed under aseptic conditions under filtered air. Living plant materials from the environment are naturally contaminated on their surfaces (and sometimes interiors) with microorganisms, so surface sterilization of starting materials (explants) in chemical solutions (usually Sodium or calcium hypochlorite or mercuric chloride) is required. Mercuric chloride is seldom used as a plant sterilant today, unless other sterilizing agents are found to be ineffective, as it is dangerous to use, and is difficult to dispose of. Explants are then usually placed on the surface of a solid culture medium, but are sometimes placed directly into a liquid medium, particularly when cell suspension cultures are desired. Solid and liquid media are generally composed of inorganic salts plus a few organic nutrients, vitamins and plant hormones. Solid media are prepared from liquid media with the addition of a gelling agent, usually purified agar.

In-vitro Tissue Culture Potato Explants

The composition of the medium, particularly the plant hormones and the nitrogen source (nitrate versus ammonium salts or amino acids) have profound effects on the morphology of the tissues that grow from the initial explant. For example, an excess of auxin will often result in a proliferation of roots, while an excess of cytokinin may yield shoots. A balance of both auxin and cytokinin will often produce an unorganised growth of cells, or callus, but the morphology of the outgrowth will depend on the plant species as well as the medium composition. As cultures grow, pieces are typically sliced off and transferred to new media (subcultured) to allow for growth or to alter the morphology of the culture. The skill and experience of the tissue culturist are important in judging which pieces to culture and which to discard. As shoots emerge from a culture, they may be sliced off and rooted with auxin to produce plantlets which, when mature, can be transferred to potting soil for further growth in the greenhouse as normal plants.

Choice of Explant

The tissue obtained from the plant to culture is called an explant. Based on work with certain model systems, particularly tobacco, it has often been claimed that a totipotent explant can be grown from any part of the plant. However, this concept has been vitiated in practice. In many species explants of various organs vary in their rates of growth and regeneration, while some do not grow at all. The choice of explant material also determines if the plantlets developed via tissue culture are haploid or diploid. Also the risk of microbial contamination is increased with inappropriate explants. Thus it is very important that an appropriate choice of explant be made prior to tissue culture. The specific differences in the regeneration potential of different organs and explants have various explanations.

The significant factors include differences in the stage of the cells in the cell cycle, the availability of or ability to transport endogenous growth regulators, and the metabolic capabilities of the cells. The most commonly used tissue explants are the meristematic ends of the plants like the stem tip, auxiliary bud tip and root tip. These tissues have high rates of cell division and either concentrate or produce required growth regulating substances including auxins and cytokinins.

The pathways through which whole plants are regenerated from cells and tissues or explants such as meristems broadly fall into three types:

1. The method in which explants that include a meristem (viz. the shoot tips or nodes) are grown on appropriate media supplemented with plant growth regulators to induce proliferation of multiple shoots, followed by rooting of the excised shoots to regenerate whole plants,
2. The method in which totipotency of cells is realised in the form of *de novo* organogenesis, either directly in the form of induction of shoot meristems on the explants or indirectly via a callus (unorganised mass of cells resulting from proliferation of cells of the explant) and plants are regenerated through induction of roots on the resultant shoots,

3. Somatic embryogenesis, in which asexual adventive embryos(comparable to zygotic embryos in their structure and development) are induced directly on explants or indirectly through a callus phase.

The first method involving the meristems and induction of multiple shoots is the preferred method for the micropropagation industry since the risks of somaclonal variation (genetic variation induced in tissue culture) are minimal when compared to the other two methods. Somatic embryogenesis is a method that has the potential to be several times higher in multiplication rates and is amenable to handling in liquid culture systems like bioreactors.

Some explants, like the root tip, are hard to isolate and are contaminated with soil microflora that become problematic during the tissue culture process. Certain soil microflora can form tight associations with the root systems, or even grow within the root. Soil particles bound to roots are difficult to remove without injury to the roots that then allows microbial attack. These associated microflora will generally overgrow the tissue culture medium before there is significant growth of plant tissue.

Aerial (above soil) explants are also rich in undesirable microflora. However, they are more easily removed from the explant by gentle rinsing, and the remainder usually can be killed by surface sterilization. Most of the surface microflora do not form tight associations with the plant tissue. Such associations can usually be found by visual inspection as a mosaic, decolourisation or localised necrosis on the surface of the explant. An alternative for obtaining uncontaminated explants is to take explants from seedlings which are aseptically grown from surface-sterilized seeds. The hard surface of the seed is less permeable to penetration of harsh surface sterilizing agents, such as hypochlorite, so the acceptable conditions of sterilization used for seeds can be much more stringent than for vegetative tissues.

Tissue cultured plants are clones, if the original mother plant used to produce the first explants is susceptible to a pathogen or environmental condition, the entire crop would be susceptible to the same problem, conversely any positive traits would remain within the line also.

Applications

Plant tissue culture is used widely in plant science; it also has a number of commercial applications. Applications include:

- Micropropagation is widely used in forestry and in floriculture. Micropropagation can also be used to conserve rare or endangered plant species.
- A plant breeder may use tissue culture to screen cells rather than plants for advantageous characters, e.g. herbicide resistance/tolerance.

- Large-scale growth of plant cells in liquid culture in bioreactors for production of valuable compounds, like plant-derived secondary metabolites and recombinant proteins used as biopharmaceuticals.
- To cross distantly related species by protoplast fusion and regeneration of the novel hybrid.
- To cross-pollinate distantly related species and then tissue culture the resulting embryo which would otherwise normally die (Embryo Rescue).
- For production of doubled monoploid (dihaploid) plants from haploid cultures to achieve homozygous lines more rapidly in breeding programmes, usually by treatment with colchicine which causes doubling of the chromosome number.
- As a tissue for transformation, followed by either short-term testing of genetic constructs or regeneration of transgenic plants.
- Certain techniques such as meristem tip culture can be used to produce clean plant material from virused stock, such as potatoes and many species of soft fruit.
- micropropagation using meristem and shoot culture to produce large numbers of identical individuals.
- production of identical sterile hybrid species can be obtained.

Laboratories

Although some growers and nurseries have their own labs for propagating plants by the technique of tissue culture, a number of independent laboratories provide custom propagation services. The Plant Tissue Culture Information Exchange lists many commercial tissue culture labs. Since plant tissue culture is a very labour intensive process, this would be an important factor in determining which plants would be commercially viable to propagate in a laboratory.

Organ Culture

Organ culture is a development from tissue culture methods of research, the organ culture is able to accurately model functions of an organ in various states and conditions by the use of the actual *in vitro* organ itself. Parts of an organ or a whole organ can be cultured in vitro. The main objective is to maintain the architecture of the tissue and direct it towards normal development. In this technique, it is essential that the tissue is never be disrupted or damaged. It thus requires careful handling. The media used for a growing organ culture are generally the same as those used for tissue culture. The techniques for organ culture can be classified into (i) those employing a solid medium and (ii) those employing liquid medium.

Methodology

Embryonic organ culture is an easier alternative to normal organ culture derived from adult animals. The following are three techniques employed for embryonic organ culture.

Plasma clot Method

The following are general steps in organ culture on plasma clots.

1. Prepare a plasma clot by mixing 15 drops of plasma with five drops of embryo extract in a watch glass.
2. Place a watch glass on a pad of cotton wool in a petri dish; cotton wool is kept moist to prevent excessive evaporation from the dish.
3. Place a small, carefully dissected piece of tissue on top of the plasma clots in watch glass.

The technique has now been modified, and a raft of lens paper or rayon net is used on which the tissue is placed. Transfer of the tissue can then be achieved by raft easily. Excessive fluid is removed and the net with the tissue placed again on the fresh pool of medium. Parts of an organ or a whole organ can be cultured in vitro. The main objective is to maintain the architecture of the tissue and direct it towards normal development. In this technique, it is essential that the tissue never be disrupted or damaged. It thus requires careful handling. The media used for a growing organ are generally the same as those used for tissue culture. The techniques for organ culture can be classified into (i) those employing a solid medium and (ii) those employing liquid medium.

Agar Gel Method

Media solidified with agar are also used for organ culture and these media consist of 7 parts 1% agar in BSS, 3 parts chick embryo extract and 3 parts of horse serum. Defined media with or without serum are also used with agar. The medium with agar provides the mechanical support for organ culture. It does not liquefy. Embryonic organs generally grow well on agar, but adult organ culture will not survive on this medium. The culture of adult organs or parts from adult animals is more difficult due to their greater requirement of oxygen. A variety of adult organs (e.g. the liver) have been cultured using special media with special apparatus (Towell's II culture chamber). Since serum was found to be toxic, serum-free media were used, and the special apparatus permitted the use of 95% oxygen.

Raft Methods

In this approach the explant is placed onto a raft of lens paper or rayon acetate, which is floated on serum in a watch glass. Rayon acetate rafts are made to float on the serum by treating their 4 corners with silicone. Similarly, floatability of lens paper is enhanced by treating it with silicone. On each raft, 4 or more explants are usually placed. In a combination of raft and clot techniques, the explants are first placed on a suitable raft, which is then kept on a plasma clot. This modification makes media changes easy, and prevents the sinking of explants into liquefied plasma.

Grid Method

Initially devised by Trowell in 1954, the grid method utilises 25 mm x 25 mm pieces of a suitable wire mesh or perforated stainless steel sheet whose edges are bent to form 4 legs of about 4 mm height. Skeletal tissues are generally placed directly on the grid but softer tissues like glands or skin are first placed on rafts, which are then kept on the grids. The grids themselves are placed in a culture chamber filled with fluid medium up to the grid; the chamber is supplied with a mixture of O_2 and CO_2 to meet the high O_2 requirements of adult mammalian organs. A modification of the original grid method is widely used to study the growth and differentiation of adult and embryonic tissues.

Limitations

- Results from organ cultures are often not comparable to those from whole animals studies, e.g. in studies on drug action since the drugs are metabolised in vivo but not in vitro.

Current Progress

In April 2006, scientists reported a successful trial of seven bladders grown in-vitro and given to humans. A bladder has been cultured by Anthony Atala of the Wake Forest Institute for Regenerative Medicine in Winston-Salem, North Carolina. A jawbone has been cultured at Columbia University, a lung has been cultured at Yale. A beating rat heart has been cultured by Doris Taylor at the University of Minnesota. An artificial kidney has been cultured by H. David Humes at the University of Michigan.

List of Animals that have been Cloned

This is a list of animals that have been cloned in alphabetical order. One significant aspect of this list is documenting the transition from early concerns that animal cloning procedures might be limited to a few species, that cloned animals might be physiologically abnormal, or cloning might lack utility for societ

Carp

Chinese embryologist Tong Dizhou successfully inserted the DNA from a male Asian carp into the egg of a female Asian carp to create the first fish clone in 1963. In 1973 Dizhou inserted Asian carp DNA into a European crucian carp to create the first interspecies clone.

Cat

- In December 2001, scientists at Texas A&M University created the first cloned cat, CC (CopyCat). Even though CC is an exact copy of his host, they have different personalities; i.e. CC is shy and timid, his host on the other hand is playful and curious.
- In 2004, the first commercially cloned cat, Little Nicky, was created by Genetic Savings & Clone.

Cattle

- First World cloned calf (Gene) was born on February 7, 1997 on American Breeders Service facilities in Deforest, Wisconsin. Later it was transferred and kept to Minnesota Zoo Education Centre.
- A Holstein heifer named Amy was cloned by Dr. Xiangzhong (Jerry) Yang using ear skin cells from a high-merit cow named Aspen at the University of Connecticut on June 10, 1999, followed by three additional clones, Betty, Cathy and Daisy by July 7, 1999.
- Second Chance, a Brahman bull was cloned from Chance, a beloved celebrity bull. Second Chance was born August 9, 1999 at Texas A&M University.
- Texas A&M University cloned a Black Angus bull named 86 Squared in 2000, after cells from his donor, Bull 86, had been frozen for 15 years. Both bulls exhibit a natural resistance to Brucellosis, Tuberculosis and other diseases which can be transferred in meat.
- Millie and Emma were two female Jersey cows cloned at the University of Tennessee in 2001. They were the first cows to be produced using standard cell-culturing techniques.
- Pampa the first animal cloned in Argentina by Biosidus (2002)
- Ten more Jersey cows were cloned at the University of Tennessee. (females, 2002)
- Bonyana and Tamina cloned calf in Royan Research Institute, Isfahan, Iran in summer of 2009.
- In 2010 the first Spanish Fighting Bull was cloned by Spanish scientists.
- Anatolian Grey bull (Efe) was cloned in Turkey in 2009 and cattle from the same breed (Ece, Ecem, Nilufer, Kiraz) by TUBITAK.
- GARIMA- I: world's first buffalo calf through the "Hand guided Cloning Technique" was born on February 6, 2009 at NDRI, Karnal(India).

- GARIMA- II: NDRI, Karnal(India).
- Cloned male buffalo calf Shresth born on August 26, 2010 at National Dairy Research Institute, Karnal, India-132001.

Deer

- Dewey was born on May 23, 2003 at Texas A&M University.

Dog

- South Korean scientist Hwang Woo-Suk cloned the first dog, an afghan hound named Snuppy. Later in 2005 Hwang Woo-Suk was found to have fabricated evidence in stem cell research projects. This caused some to question the veracity of his other experiments, including Snuppy. In their investigation of Hwang Woo-Suk's publication, however, a team from SNU confirmed that Snuppy was a true clone of Tei, the DNA donor dog. South Korean scientists recently cloned 'sniffer' dogs.
- BioArts International held a dog cloning contest where people would send in submissions about which dog was the most suited to be cloned. The winner was Trakr, a K-9 police dog who was a 9/11 hero.
- In summer 2011, South Korean researchers cloned a beagle dog named Tegon, which glowed in ultraviolet light

Ferret

Clones Libby and Lilly were produced via nuclear transfer by cell fusion in 2004.

Frog (Tadpole)

In 1958, John Gurdon, then at Oxford University, explained that he had successfully cloned a frog. He did this by using intact nuclei from somatic cells from a Xenopus tadpole. This was an important extension of work of Briggs and King in 1952 on transplanting nuclei from embryonic blastula cells

Fruit Flies

- (2004).

Gaur

A species of wild cattle, the first endangered species to be cloned. In 2001 at the Trans Ova Genetics in Sioux Centre, Iowa, USA, a cloned Gaur was born from a surrogate domestic cow mother. However, the calf died within 48 hours.

Goat

- Downen TX 63 684 (nicknamed Megan) was cloned from a top producing Boer goat born on March 29, 2001 at Texas A&M University.

- The Middle East's first and the world's fifth cloned goat, 'Hanna', has been successfully born at Royan institute in Isfahan, Iran.

  The cloned goat was developed in the surrogate uterus of a black Bakhtiari goat for 147 days and was born, Wednesday, at 1:30 a.m. through a cesarean section. She is reported to be in a good health.

  Hanna, also known as R-CAP-C1, is completely distinguished from other goats because of its white and henna-like colour. Iran's first cloned lamb, Royana, was born September 30, 2006 in Royan institute and was able to survive the post-natal complications common in cloned animals. Iranian researchers are looking to use cloned goats to produce the genetically modified animals required for manufacturing new recombinant medications.

Horse

- Prometea, female, born May 2003, Italy
- Pieraz, male, born February 2005, Italy
- Paris-Texas, male, born March 2005, USA
- Gemini, male, born September 2008, USA, clone of multiple recipient of "Horse of the Year" award for jumping Gem Twist
- Saphir, male, born February 2010, USA, clone of show jumper Sapphire.

Mice

- Possibly the first cloned mammals was a mouse (named "Masha") in 1986, in the Soviet Union. However, the cloning was done from an embryo cell, while the sheep Dolly in 1996 was cloned from an adult cell.
- The first mouse from adult cells, Cumulina, was born in 1997 at the University of Hawai'i at Mânoa in the laboratory of Ryuzo Yanagimachi using the Honolulu technique.
- Over a dozen clones as of 2002.

Mouflon

- An endangered species, the Mouflon was the first to live past infancy. Cloned 2001

Mule

- Idaho Gem (male, May 2003)
- Utah Pioneer (male, June 2003)
- Idaho Star (male, July 2003).

Pig

- 5 Scottish PPL piglets (Millie, Alexis, Dotcom, Carrel, and Christa) (March 5, 2000)
- Xena (female, August 2000).

Pyrenean Ibex

In 2009, one clone was born alive, but died seven minutes later, due to physical defects in the lungs. The Pyrenean Ibex became the first taxon ever to come back from extinction, for a period of seven minutes in January 2009.

Rabbit

- In France (March–April, 2003).

Rat

- Ralph (male, 2003).

Rhesus Monkey

- Tetra (female, January 2000) by embryo splitting.
- Cloned embryos (November 2007) by tranfer of DNA from adult cells.

Sheep

- From early embryonic cells by Steen Willadsen (1986). Megan and Morag cloned from differentiated embryonic cells in June 1995.
- Dolly (1996–2003), first cloned mammal from somatic cells
- Polly and Molly (July 1997), first transgenic cloned mammal
- Royana (2006) cloned in Royan Research institute in Isfahan, Iran.
- Oyalý and Zarife were cloned in November 2007 in Istanbul University in Istanbul, Turkey.

Water Buffalo

The world's first water buffalo was cloned either in Beijing China in 2005 or New Delhi, India in 2009 "Samrupa", the world's first cloned buffalo calf, which died a week later from a lung infection.

Wolf

- An endangered species of wolf cloned by Korean scientists including the controversial scientist Hwang Woo-Suk
- There are two cloned wolves in a zoo in Korea for public view, they are called Snuwolf and Snuwolffy which are names taken from the university in Korea, Seoul National University.

Fish Cloning

Researchers are pondering how an exclusively-female fish species has survived without genetically self-destructing itself into extinction, the BBC reports. The Amazon Molly, found in Mexico and Texas, reproduces by cloning, and while it does "interact" with chaps from other species to "trigger its reproduction process", doesn't utilise any of their DNA in producing offspring.

While this might be considered proof of the old saw that "a woman needs a man like a fish needs a bicycle", scientists from the University of Edinburgh applied "complex mathematical models" to demonstrate that the Amazon Molly should in fact have become extinct within the last 70,000 years as a result of the adverse genetic changes resulting from asexual reproduction. Quite how it has survived remains a mystery, although one theory suggests occasional liaisons with males in which DNA is taken on board to "refresh" the species' genetic make-up.

Dr Laurence Loewe, of the uni's School of Biological Sciences, said: "What we have shown now is that this fish really has something special going on and that some special tricks exist to help this fish survive. Maybe there is still occasional sex with strangers that keeps the species alive. Future research may give us some answers."

Holly, the World's First Semi-cloned Fish

Some 8 years ago, semi-cloning was considered a scientific fiction and 26 years ago, using mice for the experiment, British scientist Sir Martin Evans attempted to make haploid embryonic stem (ES) cells. Fast-forward to 2009 and it is now a team led by Assoc Prof Hong Yunhan of the Department of Biological Sciences, who have successfully produced the world's first haploid embryonic stem cells and semi-cloned fish. Their work was published in the 16 October issue of the highly acclaimed Science magazine.

"While traditional cloning uses a 'replacement' strategy, semi-cloning adopts a 'combination' strategy. In semi-cloning, scientists combine two haploid nuclei - one haploid nucleus from the embryonic stem cell mimicking the sperm nucleus, and the nucleus of an unfertilized egg (from the mother) - using nuclear transfer," said Assoc Prof Hong.

> *"In the practice of nuclear transfer for the production of Holly, we introduced a haploid embryonic stem cell nucleus into a mature egg. In this way, scientists mimic the reproduction process artificially, creating offspring that carry genetic traits from both parents similar to bisexual reproduction," he added.*

For the study, the team worked the Medaka fish as it shares a number of genes with humans in the area of reproduction. The Medaka fish is also easy to maintain being small in size and producing 30 to 40 eggs a day. Experimenting with fish also ensured that there would be no ethical issues. The findings of Assoc Prof Hong have important implications for reproductive medicine and technology. With semi-cloning which is 50 to 100 times more efficient than cloning, there is hope for the treatment of infertility, in particular male infertility which is more prominent.

In addition, the NUS team's study has generated haploid embryonic stem cells which could be used to study the effect of recessive mutations of essential genes that may not be apparent in normal diploid cells. This spells the possibility of developing mammalian haploid embryonic stem cell cultures for disease analysis. Assoc Prof Hong said that the study using the fish model took some five years and a funding of SGD1.5 million to complete. This included time for consolidation of ideas, practical and administrative work. The next step would be to conduct the study on mice which will take approximately another five years. He is also currently looking for partners in Asia for collaboration. A team of scientists at the Michigan State University has found a new, more efficient method for cloning zebra fish.

"After the mouse, it is the most commonly used vertebrate in genetic studies," said Jose Cibelli, an MSU professor of animal science and one of the paper's co-authors. *"It is used in cancer research and cardiovascular research because they have many of the same genes we have."*

Zebra fish is also used by scientists researching normal development and birth defects, as well as various human diseases and the functions of cell populations within organs. Up until now, zebra fish cloning has had a low success rate, but the new Michigan method has changed this. The new method uses ovarian fluid from a Chinook salmon to keep the unfertilized egg alive.

"This worked well, because it kept the egg inactive for some time", Cibelli explained. *It gave us two or three hours to work with.*

During the next step of the process, the Michigan researchers used a laser to remove DNA from the egg; a method borrowed from human in vitro fertilization. Next, the team devised a new, more efficient way of inserting the donor cells into the egg. Cytoplasmic Impact on Cross-Genus Cloned Fish Derived from Transgenic Common Carp (*Cyprinus carpio*) Nuclei and Goldfish (*Carassius auratus*) Enucleated Eggs.

Most cloned animals were obtained by intraspecies nuclear transfer and are phenotypically identical to their nuclear donors; furthermore, there was no further report on successful fish cloning since the report of cloned zebrafish. Here we report the production of seven cross-genus cloned fish by transferring nuclei from transgenic common carp into enucleated eggs of goldfish. Nuclear genomes of the cloned fish were exclusively derived from the nuclear donor species, common carp, whereas the mitochondrial DNA from the donor carp gradually disappeared during the development of nuclear transfer (NT) embryos. The somite development process and somite number of nuclear transplants were consistent with the recipient species, goldfish, rather than the nuclear donor species, common carp. This resulted in a long-lasting effect on the vertebral numbers of the cloned fish, which belonged to the range of goldfish. These demonstrate that fish egg cytoplasm not only can support the development driven by transplanted nuclei from a distantly related species at the genus scale but also can modulate development of the nuclear transplants.

With the advent of cloned livestock, yet another biotech science experiment may soon find its way to the American dinner table. In January 2008, the FDA essentially told the public that the meat and milk from cloned livestock are safe for human consumption. FDA's action flies in the face of widespread scientific concern about the risks of food from clones, and ignores the animal cruelty and troubling ethical concerns that the cloning process brings. The approval also goes against the will of Congress, who voted twice in 2007 to delay FDA's decision on cloned animals until additional safety and economic studies can be completed, and ignores the feelings of the American public, 150,000 of whom wrote to FDA opposing the approval during last year's public comment period. What's worse, FDA will not require labelling on cloned food, so consumers will have no way to avoid these experimental foods.

Animal cloning is a new technology with potentially severe risks for food safety. Defects in clones are common, and cloning scientists warn that even small imbalances in clones could lead to hidden food safety problems in clones' milk or meat. There are few studies on the risks of food from clones, and no long-term food safety studies have been done. Numerous opinion polls show that the majority of Americans do not want food from animal clones and are opposed to cloning on moral or ethical grounds.

The FDA's veterinary medicine advisory panel rebuked the agency in 2003 for its position, declaring that not enough research has been done to determine whether food derived from cloned animals is safe. In fact, livestock cloning raises numerous health and ethical concerns. Over 90 percent of cloning attempts fail, and cloned animals that are born have more health problems and higher mortality rates than sexually reproduced animals. Given that researchers do not understand many of the health problems that arise throughout the lifecycles of cloned animals, the FDA acted irresponsibly in assuming that the foods produced from these animals are safe for humans to eat. According to Ian Wilmut, the leader of the team of scientists that cloned the sheep Dolly, determining the health impacts of food derived from clones must be based on the animals' complete health profiles. Such studies have not been done.

Amphibian Cloning

Since the production of Dolly the sheep cloning methods for somatic cells have been thoroughly described and are becoming routine. However, the rate at which live clones are produced remains low in all mammalian species tested so far. Remarkably, irrespective of the cloning protocol or the donor-cell type, all clones display common abnormalities, particularly in the placenta. The process is also complicated by early mortality of somatic-cell clones and the founder mammalian clone, Dolly the sheep, died in February 2003 aged six years. Based on published data and on our own experience, our view is that mammalian somatic-cell cloning and the pioneer nuclear-transfer data from amphibians have much in common. We suggest that the only way to improve nuclear reprogramming is to modify the chromatin structure of somatic cells before nuclear transfer, to provide the oocyte with a chromosomal structure that is more compatible with the natural reprogramming machinery of the oocyte.

The ultimate test of whether the nucleus of a differentiated cell has undergone any irreversible functional restriction is to have that nucleus generate every other type of differentiated cell in the body. If each cell's nucleus is identical to the zygote nucleus, then each cell's nucleus should also be totipotent (capable of directing the entire development of the organism) when transplanted into an activated enucleated egg. As early as 1895, the embryologist Yves Delage predicted that "If, without deterioration, the egg nucleus could be replaced by the nucleus of an ordinary embryonic cell, we should probably see this egg developing without changes". Before such an experiment could be done, however, three techniques for transplanting nuclei into eggs had to be perfected: (1) a method for enucleating host eggs without destroying them; (2) a method for isolating intact donor nuclei; and (3) a method for transferring such nuclei into the host egg without damaging either the nucleus or the oocyte.

All three techniques were developed in the 1950s by Robert Briggs and Thomas King. First, they combined the enucleation of the host egg with its activation. When an oocyte (a developing egg cell) from the leopard frog *Rana pipiens* is pricked with a clean glass needle, the egg undergoes all the cytological and biochemical changes associated with fertilization. The internal cytoplasmic rearrangements of fertilization occur, and the completion of meiosis takes place near the animal pole of the cell. The meiotic spindle can easily be located as it pushes away the pigment granules at the animal pole, and puncturing the oocyte at this site causes the spindle and its chromosomes to flow outside the egg. The host egg is now considered both activated (the fertilization reactions necessary to initiate development have been completed) and enucleated.

The transfer of a nucleus into the egg is accomplished by disrupting a donor cell and transferring the released nucleus into the oocyte through a micropipette. Some cytoplasm accompanies the nucleus to its new home, but the ratio of donor to recipient cytoplasm is only 1:10^5, and the donor cytoplasm does not seem to affect the outcome of the experiments. In 1952, Briggs and King, using these techniques, demonstrated that blastula cell nuclei could direct the development of complete tadpoles when transferred into the cytoplasm of an activated enucleated frog egg. This procedure is called somatic nuclear transfer, or more commonly, cloning. What happens when nuclei from more advanced developmental stages are transferred into activated enucleated oocytes? King and Briggs (1956) found that whereas most blastula nuclei could produce entire tadpoles, there was a dramatic decrease in the ability of nuclei from later stages to direct development to the tadpole stage. When nuclei from the somatic cells of tailbud-stage tadpoles were used as donors, normal development did not occur.

However, nuclei from the germ cells of tailbud-stage tadpoles (which could give rise to a complete organism after fertilization) were capable of directing normal development in 40 percent of the blastulae that developed (Smith 1956). Thus, most somatic cells appeared to lose their ability to direct development as they became determined and differentiated.

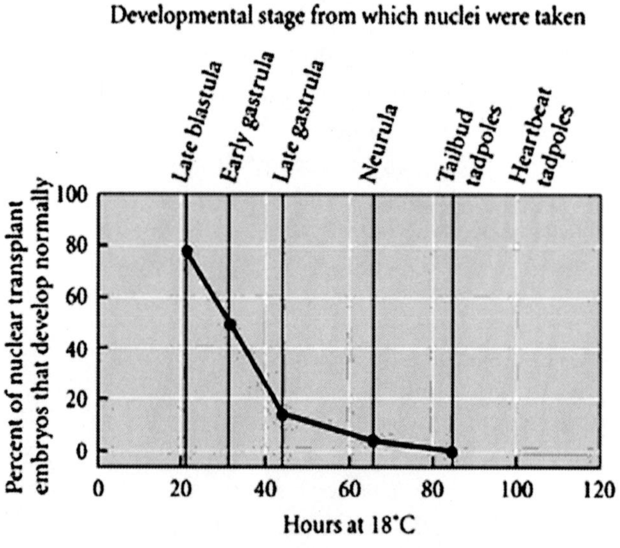

Figure: *Percentage of successful nuclear transplants as a function of the developmental age of the donor nucleus. The abscissa represents the developmental stage at which a donor nucleus (from* R. pipiens*) was isolated and inserted into an activated enucleated oocyte. The ordinate shows the percentage of those transplants capable of producing blastulae that could then direct development to the swimming tadpole stage. (After McKinnell 1978.)*

Amphibian Cloning: The Totipotency of Somatic Cells

Is it possible that some differentiated cell nuclei differ from others in their ability to direct development? John Gurdon and his colleagues, using slightly different methods of nuclear transplantation on the frog *Xenopus*, obtained results suggesting that the nuclei of some differentiated cells can remain totipotent. Gurdon, too, found a progressive loss of potency with increasing developmental age, although *Xenopus* cells retained their potencies for a longer period than did the cells of *Rana*.

To clone amphibians from the nuclei of cells known to be differentiated, Gurdon and his colleagues cultured epithelial cells from adult frog foot webbing. These cells were shown to be differentiated: each of them contained a specific keratin, the characteristic protein of adult skin cells.

When nuclei from these cells were transferred into activated enucleated *Xenopus* oocytes, none of the first-generation transfers progressed further than the formation of the neural tube shortly after gastrulation. By serial transplantation (i.e., taking nuclei from the cloned blastulas), however, numerous tadpoles were generated (Gurdon et al. 1975). Although the tadpoles all died prior to feeding, they showed that a single differentiated cell nucleus still retained incredible potencies. A nucleus of a skin cell could produce all the cells of the young tadpole.

Crmtin Diminti CMI DI CD

Chromosomal diminution (or more properly, chromatin diminution) is a fascinating exception to the general rule of the constancy of the genome. Moreover, it provides an interesting alternative to differential gene expression. Instead of repressing genes that are not being used, why not get rid of them altogether? In some species, entire chromosomes are eliminated during development (chromosome elimination).

In chromatin diminution, selected portions of the chromosomes are cast away in particular cells (Müller et al., 1996). While chromatin diminution was observed by Boveri over 100 years ago, it can now be studied on the molecular level. This has been accomplished in the roundworm *Ascaris suum* (formerly *Ascaris lumbricoides* var. *suum*), a parasite in the intestines of pigs. (Large quantities of this worm can be obtained cheaply from slaughterhouses all over the world.) These studies reveal that chromatin diminution is not a haphazard ripping apart of the genome. Rather, this diminution is a carefully orchestrated genome rearrangement, where certain regions of the DNA are degraded and new telomeres are constructed.

In *A. suum*, all the somatic cell precursors undergo chromatin diminution. The process differs slightly from the one that Boveri observed in *Parascalis univalens*. First, in *A. suum*, diminution was not seen before the third cell division. Second, there is some chromosomal fragmentation in the germline cells of *Ascaris*, and this results in the loss of some telomeric chromatin (Tobler, 1986).

Third, in *Parascaris*, about 85% of the total nuclear DNA is eliminated. In *Ascaris*, this figure is closer to 25%. (Moritz and Roth, 1976). In both species, all the heterochromatin is lost from the somatic cell nuclei. Chromatin diminution is not a general property of roundworms, nor is it in all species of parasitic roundworms.

In the free-living nematode *Caenorhabditis elegans*, there is no known chromatin diminution (Emmons et al., 1979).

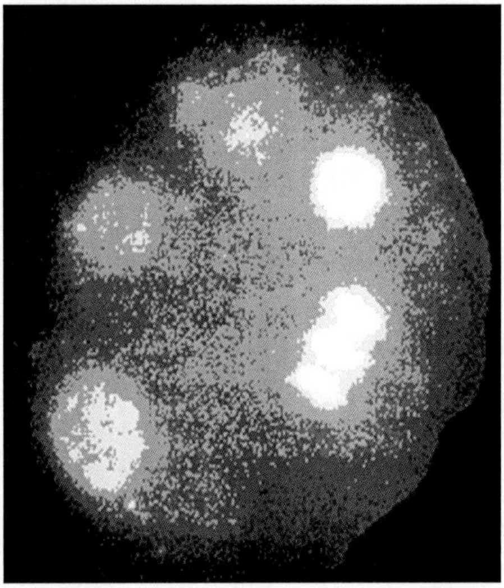

Figure : *Elimination mitosis in an early* Ascaris *embryo. The DNA is stained light blue with DAPI. One of the cells is undergoing mitosis. The eliminated chromatin is left at the equatorial plate of mitosis while the other chromatin is being drawn to the mitotic poles. (After Müller et al., 1991.)*

Chromatin diminution in *A. suum* occurs at specific chromosomal breakage regions. Nothing is known about the mechanism of this breakage or the selection of these sites.

The sites have no obvious consensus sequence. The DNA between the break site and the telomere (any thus gene in this region) is degraded. After breakage, multiple units of the telomeric sequence TTAGGC are added to the break site.

The microtubules of the spindle apparatus associate only with non-eliminated chromosomal regions. Moreover, the heteromeric, eliminated chromatin lacks kinetochore plates, thereby causing them to remain at the equatorial plate and not be drawn into the mitotic poles. Here, they are attacked by degrading enzymes and eventually disintegrate. The eliminated DNA is enriched in repetitive DNA sequences rather than in the single-copy DNA that contains most genes.

However, two single-copy genes have recently been found to become eliminated from the somatic cell precursors of *Ascaris suum*. The *fert-1* gene is on soon after fertilization, and it makes several types of transcripts.

Two of these transcripts are long-lived and persist in the somatic cells long after the genes that encode it have been eliminated (Spicher et al., 1994).

Thus, the longevity of the message may compensate for the loss of the gene.

The second gene is that encoding a germline-specific isoform of the S19 ribosomal protein. It turns out that *Ascaris* has two types of S19 genes. Both are active in the germ line, while only one is active in the somatic cells (Etter et al., 1994).

5

The Cloning Process

The most common cloning method, known as "somatic cell nuclear transfer" or simply "nuclear transfer," requires two kinds of cell. One is a somatic cell, which is collected from the animal that is to be cloned (known as the "genetic donor"). A somatic cell is any cell other than a sperm cell or egg cell, and contains the complete DNA, or genetic blueprint, of the animal it came from. For cloning purposes, somatic cells are typically obtained by a routine skin biopsy performed by a veterinarian.

The other kind of cell required for cloning is an egg cell, which is collected from a female of the same species (known as the "egg donor"). In the lab, a scientist extracts and discards the nucleus of the egg cell, which is the part of the cell that contains the egg donor's genes. The scientist then inserts the somatic cell from the genetic donor into the egg and "fuses" the two with electricity. The resulting fused egg contains the genetic donor's DNA.

The scientist stimulates the fused egg, which "activates" the egg and causes it to divide just as an egg would if it had been fertilized by a sperm cell in conventional reproduction. The activated egg is then placed in a culture medium.

As cellular division continues over the course of several days, a blastocyst (early-stage embryo) forms. After about a week, an embryo transfer specialist transfers the blastocyst to a recipient female (sometimes referred to as "surrogate mother") where it continues to develop. After a full-term pregnancy, the recipient gives birth to an animal that is essentially the identical twin of the genetic donor.

Genetics

Genetics a discipline of biology, is the science of genes, heredity, and variation in living organisms. Genetics deals with the molecular structure and function of genes, with gene behaviour in the context of a cell or organism (e.g. dominance and epigenetics), with patterns of inheritance from parent to offspring, and with gene distribution, variation and change in populations. Given that genes are universal to living organisms, genetics can be applied to the study of all living systems, from viruses and bacteria, through plants (especially crops) and domestic animals, to humans (as in medical genetics).

The fact that living things inherit traits from their parents has been used since prehistoric times to improve crop plants and animals through selective breeding. However, the modern science of genetics, which seeks to understand the process of inheritance, only began with the work of Gregor Mendel in the mid-19th century. Although he did not know the physical basis for heredity, Mendel observed that organisms inherit traits via discrete units of inheritance, which are now called genes. Genes correspond to regions within DNA, a molecule composed of a chain of four different types of nucleotides—the sequence of these nucleotides is the genetic information organisms inherit. DNA naturally occurs in a double stranded form, with nucleotides on each strand complementary to each other. Each strand can act as a template for creating a new partner strand. This is the physical method for making copies of genes that can be inherited.

The sequence of nucleotides in a gene is translated by cells to produce a chain of amino acids, creating proteins—the order of amino acids in a protein corresponds to the order of nucleotides in the gene. This relationship between nucleotide sequence and amino acid sequence is known as the genetic code. The amino acids in a protein determine how it folds into a three-dimensional shape; this structure is, in turn, responsible for the protein's function. Proteins carry out almost all the functions needed for cells to live. A change to the DNA in a gene can change a protein's amino acids, changing its shape and function: this can have a dramatic effect in the cell and on the organism as a whole. Although genetics plays a large role in the appearance and behaviour of organisms, it is the combination of genetics with what an organism experiences that determines the ultimate outcome. For example, while genes play a role in determining an organism's size, the nutrition and health it experiences after inception also have a large effect.

DNA, the molecular basis for inheritance. Each strand of DNA is a chain of nucleotides, matching each other in the centre to form what look like rungs on a twisted ladder. Although the science of genetics began with the applied and theoretical work of Gregor Mendel in the mid-19th century, other theories of inheritance preceded Mendel. A popular theory during Mendel's time was the concept of blending inheritance: the idea that individuals inherit a smooth blend of traits from their parents.

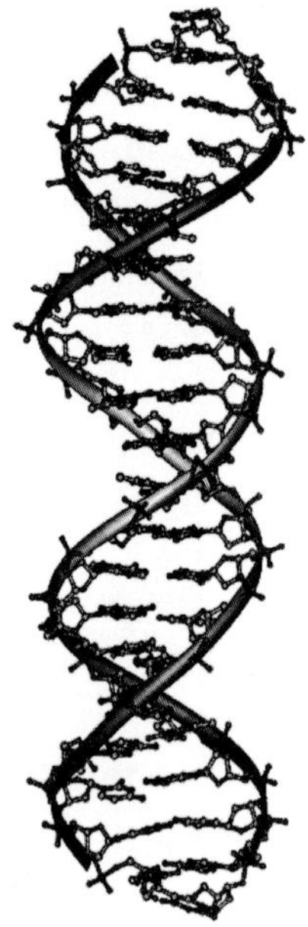

Figure : *DNA, the molecular basis for inheritance. Each strand of DNA is a chain of nucleotides, matching each other in the center to form what look like rungs on a twisted ladder.*

Mendel's work disproved this, showing that traits are composed of combinations of distinct genes rather than a continuous blend. Another theory that had some support at that time was the inheritance of acquired characteristics: the belief that individuals inherit traits strengthened by their parents. This theory (commonly associated with Jean-Baptiste Lamarck) is now known to be wrong—the experiences of individuals do not affect the genes they pass to their children. Other theories included the pangenesis of Charles Darwin (which had both acquired and inherited aspects) and Francis Galton's reformulation of pangenesis as both particulate and inherited.

Modern genetics started with Gregor Johann Mendel, a German-Czech Augustinian monk and scientist who studied the nature of inheritance in plants. In his paper "Versuche fibre Pflanzenhybriden" ("Experiments on Plant Hybridisation"), presented in 1865 to the *Naturforschender Verein* (Society for Research in Nature) in Brünn, Mendel traced the inheritance patterns of certain traits in pea plants and described them mathematically. Although this theoretically never recombine, however, were it not for the process of chromosomal crossover. During crossover, chromosomes exchange stretches of DNA, effectively shuffling the gene alleles between the chromosomes. This process of chromosomal crossover generally occurs during meiosis, a series of cell divisions that creates haploid cells. The probability of chromosomal crossover occurring between two given points on the chromosome is related to the distance between the points.

For an arbitrarily long distance, the probability of crossover is high enough that the inheritance of the genes is effectively uncorrelated. For genes that are closer together, however, the lower probability of crossover means that the genes demonstrate genetic linkage—alleles for the two genes tend to be inherited together.

The amounts of linkage between a series of genes can be combined to form a linear linkage map that roughly describes the arrangement of the genes along the chromosome.

Genes Express and their Functions

Genes generally express their functional effect through the production of proteins, which are complex molecules responsible for most functions in the cell. Proteins are made up of one or more polypeptide chains, each of which is composed of a sequence of amino acids, and the DNA sequence of a gene (through an RNA intermediate) is used to produce a specific amino acid sequence. This process begins with the production of an RNA molecule with a sequence matching the gene's DNA sequence, a process called transcription.

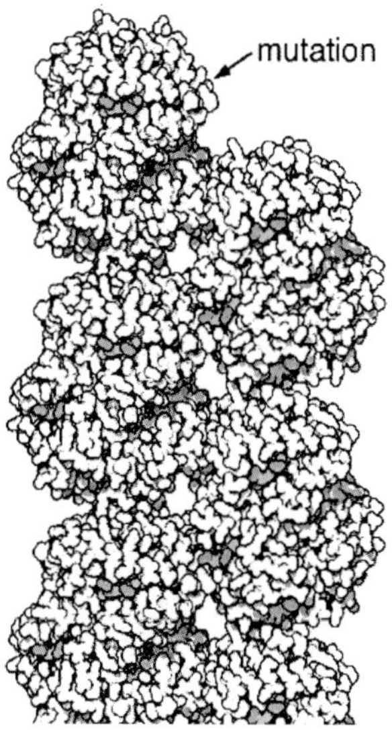

Figure : *A single amino acid change causes hemoglobin to form fibers.*

The messenger RNA molecule is then used to produce a corresponding amino acid sequence through a process called translation. Each group of three nucleotides in the sequence, called a codon, corresponds either to one of the twenty possible amino acids in a protein or an instruction to end the amino acic sequence; this correspondence is called the genetic code.

The flow of information is unidirectional: information is transferred from nucleotide sequences into the amino acid sequence of proteins, but it never transfers from protein back into the sequence of DNA—a phenomenon Francis Crick called the central dogma of molecular biology. The specific sequence of amino acids results in a unique three-dimensional structure for that protein, and the three-dimensional structures of proteins are related to their functions. Some are simple structural molecules, like the fibres formed by the protein collagen.

Proteins can bind to other proteins and simple molecules, sometimes acting as enzymes by facilitating chemical reactions within the bound molecules (without changing the structure of the protein itself). Protein structure is dynamic; the protein hemoglobin bends into slightly different forms as it facilitates the capture, transport, and release of oxygen molecules within mammalian blood. A single nucleotide difference within DNA can cause a change in the amino acid sequence of a protein.

Because protein structures are the result of their amino acid sequences, some changes can dramatically change the properties of a protein by destabilizing the structure or changing the surface of the protein in a way that changes its interaction with other proteins and molecules.

For example, sickle-cell anemia is a human genetic disease that results from a single base difference within the coding region for the â-globin section of hemoglobin, causing a single amino acid change that changes hemoglobin's physical properties.

Sickle-cell versions of hemoglobin stick to themselves, stacking to form fibres that distort the shape of red blood cells carrying the protein. These sickle-shaped cells no longer flow smoothly through blood vessels, having a tendency to clog or degrade, causing the medical problems associated with this disease.

Some genes are transcribed into RNA but are not translated into protein products—such RNA molecules are called non-coding RNA. In some cases, these products fold into structures which are involved in critical cell functions (e.g. ribosomal RNA and transfer RNA). RNA can also have regulatory effect through hybridisation interactions with other RNA molecules (e.g. microRNA).

Genetic Disorders

A genetic disorder is an illness caused by abnormalities in genes or chromosomes, especially a condition that is present from before birth. Most genetic disorders are quite rare and affect one person in every several thousands or millions. A genetic disorder may or may not be a heritable disorder. Some genetic disorders are passed down from the parents' genes, but others are always or almost always caused by new mutations or changes to the DNA. In other cases, the same disease, such as some forms of cancer, may be caused by an inherited genetic condition in some people, by new mutations in other people, and by non-genetic causes in still other people. Some types of recessive gene disorders confer an advantage in the heterozygous state in certain environments.

Single Gene Disorder

A single gene disorder is the result of a single mutated gene. There are estimated to be over 4000 human diseases caused by single gene defects. Single gene disorders can be passed on to subsequent generations in several ways. Genomic imprinting and uniparental disomy, however, may affect inheritance patterns. The divisions between recessive and dominant types are not "hard and fast" although the divisions between autosomal and X-linked types are (since the latter types are distinguished purely based on the chromosomal location of the gene).

For example, achondroplasia is typically considered a dominant disorder, but children with two genes for achondroplasia have a severe skeletal disorder that achondroplasics could be viewed as carriers of. Sickle-cell anemia is also considered a recessive condition, but heterozygous carriers have increased resistance to malaria in early childhood, which could be described as a related dominant condition. When a couple where one partner or both are sufferers or carriers of a single gene disorder and wish to have a child they can do so through IVF which means they can then have PGD (pre-implantation genetic diagnosis) to check whether the fertilized egg has had the genetic disorder passed on.

Autosomal Dominant

Only one mutated copy of the gene will be necessary for a person to be affected by an autosomal dominant disorder. Each affected person usually has one affected parent. There is a 50% chance that a child will inherit the mutated gene. Conditions that are autosomal dominant sometimes have reduced penetrance, which means that although only one mutated copy is needed, not all individuals who inherit that mutation go on to develop the disease. Examples of this type of disorder are Huntington's disease, neurofibromatosis type 1, neurofibromatosis type 2, Marfan syndrome, hereditary nonpolyposis colourectal cancer, and hereditary multiple exostoses, which is a highly penetrant autosomal dominant disorder. Birth defects are also called congenital anomalies.

Dominance (Genetics)

Dominance in genetics is a relationship between two variant forms (alleles) of a single gene, in which one allele masks the effect of the other in influencing some trait. In the simplest case, if a gene exists in two allelic forms (*A* & *a*), three combinations of alleles (genotypes) are possible: *AA*, *Aa*, and *aa*. If *Aa* individuals (heterozygotes) show the same form of the trait (phenotype) as *AA* individuals (homozygotes), and *aa* homozygotes show a different phenotype, allele *A* is said to *dominate* or *be dominant to* allele *a*, and *a* is said to *be recessive to A*.

Background: Diploid, Chromosomes, Genes, Loci, & Alleles

Diploid / Haploid

Most familiar plants, like peas, and familiar animals, like fruit flies and humans, have paired chromosomes, and are described as diploid. One chromosome of each pair is contributed by each parent: one by the female parent in her ovum, and one by the male parent in his sperm, which are joined at fertilization. The ovum and sperm cells have only one copy of each chromosome and are described as haploid. Production of haploid gametes occurs through a process called meiosis.

Chromosomes, Genes, and Alleles

Each chromosome of a matching pair is structurally similar to the other, and each member of a homologous pair has the same genetic material arranged in the same order and physical locations (loci, singular locus). The genetic material in each chromosome comprises a series of discrete genes that influence various traits. Thus, each gene also has a corresponding homologue, which may exist in different forms: the variant forms are called alleles. The alleles at the same locus on the two homologous chromosomes may be identical or different.

In popular usage, "gene" and "allele" are often used interchangeably. This produces misunderstandings. Properly, "gene" refers to a hereditary unit, ordinarily at a fixed position on a chromosome, that influences a particular trait. Genes are now understood to comprise DNA. "Allele" refers to any of the many particular forms of a gene that may be present in a population of individuals from a particular species, at a particular locus. *E.g.*, it is inaccurate to say "This pea plant has a pair of wrinkled genes", and it is more accurate to say, "This plant has two 'w' alleles for the 'Seed Shape' gene, and will produce wrinkled peas." Consider also the example of blood type in humans. Near the long arm of chromosome nine appears a gene that determines whether an individual will be blood type, A, B, or O. There are three different alleles that could be present at this locus, but only two can be present in any individual, one inherited from their mother and one from their father.

Homozygous, Heterozygous

If two alleles of a given gene are identical, the organism is called a homozygote and is homozygous with respect to that gene; if instead the two alleles are different, the organism is a heterozygote and is heterozygous. The genetic makeup of an organism, either at a single locus or over all its genes collectively, is called the genotype. The genotype of an organism directly or indirectly affects its molecular, physical and other traits, which individually or collectively are called the phenotype. At heterozygous gene loci, the two alleles interact to produce the phenotype. The simplest form of allele interaction is the one described by Mendel, now called Mendelian, in which the appearance/ phenotype caused by one allele is apparent, called dominant, and the appearance/ phenotype caused by the other allele is not apparent, called recessive.

In the simplest case, the phenotypic effect of one allele completely masks the other in heterozygous combination; that is, the phenotype produced by the two alleles in heterozygous combination is identical to that produced by one of the two homozygous genotypes. The allele that masks the other is said to be *dominant* to the latter, and the alternative allele is said to be *recessive* to the former.

Which Trait is Dominant?

The terms *dominant* and *recessive* refer to the interaction of alleles in producing the phenotype of the heterozygote. If there are two alternative phenotypes, by definition the phenotype exhibited by the heterozygote is called "dominant" and the "hidden" phenotype is called "recessive". The key concept of dominance is that the heterozygote is phenotypically *identical* to one of the two homozygotes. That trait corresponding to the dominant allele may then be called the "dominant" trait.

Dominance is a *genotypic* relationship between *alleles*, as manifested in the phenotype. It is unrelated to the nature of the phenotype itself, e.g., whether it is regarded as normal or abnormal, standard or nonstandard, healthy or diseased, stronger or weaker, or more or less extreme. It is also important to distinguish between the "round" gene locus, the "round" allele at that locus, and the "round" phenotype it produces. It is inaccurate to say that "the round gene dominates the wrinkled gene" or that "round peas dominate wrinkled peas."

Nomenclature

In genetics, the common convention is that dominant alleles are written as capital letters and recessive alleles as lower-case letters. In the pea example, once the dominance relationships of the two alleles are known, it is possible to designate the dominant allele that produces a round shape by a capital-letter symbol R, and the alternative recessive allele that produces a wrinkled shape by a lower-case symbol r. The homozygous dominant, heterozygous, and homozygous recessive genotypes are then written RR, Rr, and rr, respectively. It would also be possible to designate the two alleles as W and w, and the three genotypes WW, Ww, and ww, the first two of which produced round peas and the third wrinkled peas. Note that the choice or "R" or "W" as the symbol for the dominant allele does not prejudge whether the allele causing the "round" or "wrinkled" phenotype when homozygous is the dominant one.

Another system of notation designates the gene involved in seed shape as the "*Shp*" gene, which exists in two allelic forms, Shp^R and Shp^w, the dominance relationships of the two being indicated by the case of the superscripts. This system is the standard system in *Drosophila* genetics.

Relationship to other Genetic Concepts

The concept of dominance is involved with a number of other genetic concepts.

Multiple Alleles

Although any individual has at most two different alleles, most genes exist in a large number of allelic forms in the population as a whole. In some cases, the alleles have different effects on the phenotype, and their dominance interactions with each other can be described as a series. For example, the best known human blood groups, the ABO system, comprises three sets of alleles at the *I* locus, I^A, I^B, and I^O. The first two are dominant to the latter: that is, the AA and AO genotypes produce indistinguishable blood group phenotypes, called "*Type A*", as do BB and BO, which produce "*Type B*" blood. In another example, coat colour in siamese cats and related breeds is determined by a series of alleles at the albino gene locus (*c*) that produce different levels of pigment and hence different levels of colour dilution. Four of these are c^+, c^b, c^s, and c^a (standard, Burmese, siamese, and albino, respectively), where the first allele is completely dominant to the last three, and the last is completely recessive to the first three.

Incomplete and Semi-dominance

Complete dominance occurs when the phenotype of the heterozygote is completely indistinguishable from that of the dominant homozygote. This is frequently not the case. Incomplete dominance occurs when the phenotype of the heterozygous genotype is an intermediate of the phenotypes of the homozygous genotypes. For example, the snapdragon flower colour is either homozygous for red or white. When the red homozygous flower is paired with the white homozygous flower, the result yields a pink snapdragon flower. The pink snapdragon is the result of incomplete dominance.

Co-dominance

Co-dominance occurs when the contributions of both alleles are visible in the phenotype. In the ABO example, the I^A and I^B alleles are co-dominant in producing the AB blood group phenotype, in which both A-type and B-type antigens are made. Another example occurs at the locus for the Beta-globin component of hemoglobin, where the three molecular phenotypes of Hb^A/Hb^A, Hb^A/Hb^S, and Hb^S/Hb^S are all equally detectable by protein electrophoresis. (The medical condition produced by the heterozygous genotype is called an *incomplete dominant*). For most gene loci at the molecular level, both alleles are expressed co-dominantly, because both are transcribed into RNA.

Co-dominance is different from incomplete or semi-dominance. For example, pink flowers might be the product of two alleles that produce red and white pigments that become mixed (co-dominance on the pigment level, no dominance on the colour level), or the result of one allele that produces the usual amount of red pigment and another non-functional allele that produces no pigment, so as to produce a dilute, intermediate pink colour (no dominance at either level).

Autosomal Versus Sex-linked Dominance

In humans and other mammal species, sex is determined by two sex chromosomes called the X chromosome and the Y chromosome. Human females are typically XX; males are typically XY. The remaining pairs of chromosome are found in both sexes and are called autosomes; genetic traits due to loci on these chromosomes are described as autosomal, and may be dominant or recessive. Genetic traits on the X and Y chromosomes are called sex linked, because they tend to be characteristic of one sex or the other. In practice, the term almost always refers to X-linked traits. Females have two copies of every gene locus found on the X chromosome, just as for the autosomes, and the same dominance relationships apply. Males however have only one copy of each X chromosome gene locus, and are described as hemizygous for these genes. The

Y chromosome is much smaller than the X, and contains a much smaller set of genes that influence 'maleness', such as the SRY gene for testis determining factor.

Dominance rules for sex-linked gene loci are determined by their behaviour in the female: because the male has only one allele, that allele is always expressed regardless of whether it is dominant or recessive. An example of sex-linked is: Hair loss is a trait caused by a gene located on the X chromosomes.

Epistasis

Epistasis ["*epi* + *stasis* = to sit on top"] is an interaction between genotypes at two *different* gene loci, which sometimes resembles a dominance interaction at a single locus. Epistasis modifies the characteristic 9:3:3:1 ratio expected for two non-epistatic genes. Most genetic systems involve complex epistatic interactions among multiple gene loci.

For two loci, 14 classes of epistatic interactions are recognised. As an example of *recessive epistasis*, one gene locus may determine whether a flower pigment is yellow (AA or Aa) or green (aa), while another locus determines whether the pigment is produced (BB or Bb) or not (bb). In a bb plant, the flowers will be white, irrespective of the genotype of the other locus as AA, Aa, or aa. The b allele is *not* dominant to the A allele: the B locus shows *recessive epistasis* to the A locus, because the B locus when homozygous for the recessive allele (bb) suppresses phenotypic expression of the A locus. In a cross between two AaBb plants, this produces a characteristic 9:3:4 ratio, in this case of yellow : green : white flowers.

In *dominant epistasis*, one gene locus may determine yellow and green pigment as in the previous example: AA and Aa are yellow, and aa are green. A second locus determines whether a pigment precursor is produced (dd) or not (DD or Dd). Here, in a D- plant, the flowers will be colourless irrespective of the genotype at the *A* locus, because of the epistatic effect of the dominant D allele. Thus, in a cross between two AaDd plants, 3/4 of the plants will be colourless, and the yellow and green phenotypes are expressed only in dd plants. This produces a characteristic 12:3:1 ratio of white : yellow : green plants.

Supplementary epistasis occurs when two loci affect the same phenotype. For example, if pigment colour is produced by CC or Cc but not cc, and by DD or Dd but not dd, then pigment is produced only in C-D- genotypes, and not in any genotype combination with cc *or* dd. That is, *both* loci must have at least one dominant allele to produce the phenotype. This produces a characteristic ratio 9:7 ratio of pigmented to unpigmented plants.

Classification by Fitness or Trait Value

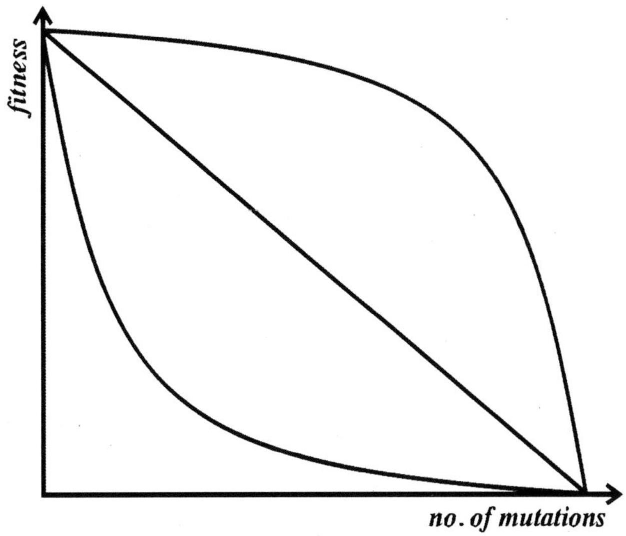

Figure: *Diagram illustrating different relationships between numbers of mutations and fitness.* Synergistic *epistasis is the blue line - each mutation has a disproportionately large effect on the organism's fitness.* Antagonistic *epistasis is the red line.*

Two-locus epistatic interactions can be either synergistic (enhancing the effectiveness) or antagonistic (reducing the activity). In the example of a haploid organism with genotypes (at two loci) *AB*, *Ab*, *aB* or *ab*, we can think of the following trait values where higher values suggest greater expression of the characteristic (the exact values are simply given as examples): Understanding whether the majority of genetic interactions are synergistic or antagonistic will help solve such problems as the evolution of sex.

Epistasis and Sex

Negative epistasis and sex are thought to be intimately correlated. Experimentally, this idea has been tested in using digital simulations of asexual and sexual populations.

Over time, sexual populations move towards more negative epistasis, or the lowering of fitness by two interacting alleles. It is thought that negative epistasis allows individuals carrying the interacting deleterious mutations to be removed from the populations efficiently. This removes those alleles from the population, resulting in an overall more fit population. This hypothesis was proposed by Alexey Kondrashov, and is sometimes known as the *deterministic mutation hypothesis* and has also been tested using artificial gene networks.

However, the evidence for this hypothesis has not always been straightforward and the model proposed by Kondrashov has been criticized for assuming mutation parameters far from real world observations. In addition, in those tests which used artificial gene networks, negative epistasis is only found in more densely connected networks, whereas empirical evidence indicates that natural gene networks are sparsely connected, and theory shows that selection for robustness will favour more sparsely connected and minimally complex networks.

Functional or Mechanistic Classification

- Genetic suppression - the double mutant has a less severe phenotype than either single mutant. [This term can also apply to a case where the double mutant has a phenotype intermediate between those of the single mutants, in which case the more severe single mutant phenotype is "suppressed" by the other mutation or genetic condition. For example, in a diploid organism, a hypomorphic (or partial loss-of-function) mutant phenotype can be suppressed by knocking out one copy of a gene that acts oppositely in the same pathway. In this case, the second gene is described as a "dominant suppressor" of the hypomorphic mutant; "dominant" because the effect is seen when one wild-type copy of the suppressor gene is present. For most genes, the phenotype of the heterozygous suppressor mutation by itself would be wild type (because most genes are not haplo-insufficient), so that the double mutant (suppressed) phenotype is intermediate between those of the single mutants.]
- Genetic enhancement - the double mutant has a more severe phenotype than one predicted by the additive effects of the single mutants.
- Synthetic lethality or unlinked non-complementation - two mutations fail to complement and yet do not map to the same locus.
- Intragenic complementation, allelic complementation, or interallelic complementation - two mutations map to the same locus, yet the two alleles complement in the heteroallelic diploid. Causes of intragenic complementation include:
    - o homology effects such as transvection, where, for example, an enhancer from one allele acts in *trans* to activate transcription from the promoter of the second allele.
    - o trans-splicing of two mutant RNA molecules to produce a functional RNA.
    - o At the protein level, another possibility involves proteins that normally function as dimers. In a heteroallelic diploid, two different

abnormal proteins could form a functional dimer if each can compensate for the lack of function in the other.

Molecular Mechanisms

The molecular basis of dominance was unknown to Mendel. It is now understood that a gene locus includes a long series (hundreds to thousands) of bases or nucleotides of deoxyribonucleic acid (DNA) at a particular point on a chromosome. The central dogma of molecular biology states that "*DNA makes RNA makes protein*", that is, that DNA is transcribed to make an RNA copy, and RNA is translated to make a protein. In this process, different alleles at a locus may or may not be transcribed, and if transcribed may be translated to slightly different forms of the same protein (called isoforms).

Proteins often function as enzymes that catalyse chemical reactions in the cell, which directly or indirectly produce phenotypes. In any diploid organism, the DNA sequences of the two alleles present at any gene locus may be identical (homozygous) or different (heterozygous). Even if the gene locus is heterozygous at the level of the DNA sequence, the proteins made by each allele may be identical. In the absence of any difference between the protein products, neither allele can be said to be dominant. Even if the two protein products are slightly different (allozymes), it is likely that they produce the same phenotype with respect to enzyme action, and again neither allele can be said to be dominant.

Dominance typically occurs when one of the two alleles is non-functional at the molecular level, that is, it is not transcribed or else does not produce a protein product. This can be the result of a mutation that alters the DNA sequence of the allele. An organism homozygous for the non-functional allele will generally show a distinctive phenotype, due to the absence of the protein product. For example, in humans and other organisms, the unpigmented skin of the albino phenotype results when an individual is homozygous for an allele that prevents synthesis of the skin pigment protein melanin. It is important to understand that it is not the lack of function that allows the allele to be described as recessive: this is the interaction with the alternative allele in the heterozygote. Three general types of interaction are possible:

In the typical case, the single functional allele makes sufficient protein to produce a phenotype identical to that of the homozygote: this is called *haplosufficiency*. For example, suppose the standard amount of enzyme produced in the functional homozygote is 100%, with the two functional alleles contributing 50% each. The single functional allele in the heterozygote produces 50% of the standard amount of enzyme, which is sufficient to produce the standard phenotype. If the heterozygote and the functional-allele homozygote have identical phenotypes, the functional allele is dominant to the non-functional allele. This occurs at the albino gene locus: the heterozygote produces sufficient enzyme to convert the pigment precursor to melanin, and the individual has standard pigmentation.

Alternatively, a single functional allele in the heterozygote may produce insufficient gene product for proper function, and the phenotype resembles that of the homozygote for the non-functional allele. This *haploinsufficiency* is much less common: usually the deficiency of gene product results in *incomplete dominance.*

The intermediate interaction occurs where the heterozygous genotype produces a phenotype intermediate between the two homozygotes. Depending on which of the two homozygotes the heterozygote most resembles, one allele is said to show *incomplete dominance* over the other. For example, in humans the *Hb* gene locus is responsible for the Beta-chain protein (HBB) that is one of the two globin proteins that make up the blood pigment hemoglobin Many people are homozygous for an allele called Hb^A; some persons carry an alternative allele called Hb^S, either as homozygotes or heterozygotes.

The hemoglobin molecules of Hb^S/Hb^S homozygotes undergo a change in shape that distorts the morphology of the red blood cells, and causes a severe, life-threatening form of anemia called sickle-cell anemia. Persons heterozygous Hb^A/Hb^S for this allele have a much less severe form of anemia called sickle-cell trait. Because the disease phenotype of Hb^A/Hb^S heterozygotes is more similar to but not identical to the Hb^A/Hb^A homozygote, the Hb^A allele is said to be *incompletely dominant* to the Hb^S allele.

In some cases, dominance of a non-standard allele results when that allele produces a defective protein that interferes with the proper function of the protein produced by the standard allele. The presence of the defective protein "dominates" the standard protein, and the disease phenotype of the heterozygote more closely resembles that of the homozygote for two variant alleles. This phenomenon occurs in a number of trinucleotide repeat diseases: for an example and more details see Huntington Disease.

Dominant and Recessive Genetic Diseases in Humans

In humans, many genetic traits or diseases are classified simply as "dominant" or "recessive." Especially with respect to so-called recessive diseases, this can oversimplify the underlying molecular basis and lead to misunderstanding of the nature of dominance. For example, the genetic disease phenylketonuria (PKU) results from any of a large number (>60) of alleles at the gene locus for the enzyme phenylalanine hydroxylase (PAH). Many of these alleles produce little or no PAH, as a result of which the substrate phenylalanine and its metabolic byproducts accumulate in the central nervous system and can cause severe mental retardation if untreated.

The genotypes and phenotypic consequences of interactions among three alleles are shown in the following table:

Genotype	*PAH activity*	*[phe] conc*	*PKU ?*
AA	100%	60 uM	No
AB	30%	120 uM	No
CC	5%	200 ~ 300 uM	Hyperphenylalanemia
BB	0.3%	600 ~ 2400 uM	Yes

In unaffected persons homozygous for a standard functional allele (AA), PAH activity is standard (100%), and the concentration of phenylalanine in the blood [phe] is about 60 uM. In untreated persons homozygous for one of the PKU alleles (BB), PAH activity is close to zero, [phe] ten to forty times standard, and the individual manifests PKU.

In the AB heterozygote, PAH activity is only 30% (not 50%) of standard, blood [phe] is elevated two-fold, and the person does not manifest PKU. Thus, the A allele is dominant to the B allele with respect to PKU, but the B allele is incompletely dominant to the A allele with respect to its molecular effect, determination of PAH activity level (0.3% < 30% << 100%). Finally, the A allele is an incomplete dominant to B with respect to [phe], as 60 uM < 120 uM << 600 uM. Note once more that it is irrelevant to the question of dominance that the recessive allele produces a more extreme [phe] phenotype.

For a third allele C, a CC homozygote produces a very small amount of PAH enzyme, which results in a somewhat elevated level of [phe] in the blood, a condition called hyperphenylalanemia, which does not result in mental retardation. That is, the dominance relationships of any two alleles may vary according to which aspect of the phenotype is under consideration. It is typically more useful to talk about the phenotypic consequences of the allelic interactions involved in any genotype, rather than to try to force them into dominant and recessive categories.

History

The concept of dominance was first described by the "Father of Genetics," Gregor Mendel, in the 1860s. Mendel observed that, for a variety of traits of garden peas having to do with the appearance of seeds, seed pods, and plant appearance, there occurred two discrete phenotypes: round *vs* wrinkled, or yellow *vs* green seeds, red *vs* white flowers, tall *vs* short plants, and so on.

When bred separately, the plants always produced the same phenotypes, generation after generation. However, when lines with different phenotypes were crossed (interbred), one and only one of the parental phenotypes showed up in the offspring: green, or round, or red, or tall, and so on. However, when these hybrid plants were crossed, the offspring plants showed the two original phenotypes, in a characteristic 3:1 ratio, with the more common type having the phenotype of the parental hybrid plants.

Mendel reasoned that each of the parents in the first cross were homozygotes for different alleles (AA and aa), that each contributed one allele to the offspring, such that all of these hybrids were heterozygotes (Aa), and that one of the two alleles in the hybrid cross *dominated* expression of the other: A masked a. The final cross between two heterozygotes (Aa X Aa) would produce AA, Aa, and aa offspring in a 1:2:1 *genotype* ratio with the first three classes showing the "A" phenotype, and the last showing the "a" phenotype, thereby producing the 3:1 *phenotype* ratio.

Mendel did not use the terms gene, allele, phenotype, genotype, homozygote, and heterozygote, all of which were introduced afterward. He did introduce the notation of capital and lowercase letters for dominant and recessive alleles, respectively, still in use today.

Autosomal Recessive

Two copies of the gene must be mutated for a person to be affected by an autosomal recessive disorder. An affected person usually has unaffected parents who each carry a single copy of the mutated gene (and are referred to as carriers). Two unaffected people who each carry one copy of the mutated gene have a 25% chance with each pregnancy of having a child affected by the disorder. Examples of this type of disorder are cystic fibrosis, sickle-cell disease, Tay-Sachs disease, Niemann-Pick disease, spinal muscular atrophy, and Roberts syndrome. Certain other phenotypes, such as wet versus dry earwax, are also determined in an autosomal recessive fashion.

X-linked Dominant

X-linked dominant disorders are caused by mutations in genes on the X chromosome. Only a few disorders have this inheritance pattern, with a prime example being X-linked hypophosphatemic rickets. Males and females are both affected in these disorders, with males typically being more severely affected than females. Some X-linked dominant conditions such as Rett syndrome, incontinentia pigmenti type 2 and Aicardi syndrome are usually fatal in males either in utero or shortly after birth, and are therefore predominantly seen in females. Exceptions to this finding are extremely rare cases in which boys with Klinefelter syndrome (47,XXY) also inherit an X-linked dominant condition and exhibit symptoms more similar to those of a female in terms of disease severity.

The chance of passing on an X-linked dominant disorder differs between men and women. The sons of a man with an X-linked dominant disorder will all be unaffected (since they receive their father's Y chromosome), and his daughters will all inherit the condition.

A woman with an X-linked dominant disorder has a 50% chance of having an affected fetus with each pregnancy, although it should be noted that in cases such as incontinentia pigmenti only female offspring are generally viable. In addition, although these conditions do not alter fertility per se, individuals with Rett syndrome or Aicardi syndrome rarely reproduce.

X-linked Recessive

X-linked recessive conditions are also caused by mutations in genes on the X chromosome. Males are more frequently affected than females, and the chance of passing on the disorder differs between men and women. The sons of a man with an X-linked recessive disorder will not be affected, and his daughters will carry one copy of the mutated gene.

A woman who is a carrier of an X-linked recessive disorder (X^RX^r) has a 50% chance of having sons who are affected and a 50% chance of having daughters who carry one copy of the mutated gene and are therefore carriers. X-linked recessive conditions include the serious diseases Hemophilia A, Duchenne muscular dystrophy, and Lesch-Nyhan syndrome as well as common and less serious conditions such as male pattern baldness and red-green colour blindness. X-linked recessive conditions can sometimes manifest in females due to skewed X-inactivation or monosomy X (Turner syndrome).

Y-linked

Y-linked disorders are caused by mutations on the Y chromosome. Because males inherit a Y chromosome from their fathers, *every* son of an affected father will be affected. Because females inherit an X chromosome from their fathers, female offspring of affected fathers are *never* affected. Since the Y chromosome is relatively small and contains very few genes, there are relatively few Y-linked disorders. Often the symptoms include infertility, which may be circumvented with the help of some fertility treatments. Examples are male infertility and hypertrichosis pinnae.

Mitochondrial

This type of inheritance, also known as maternal inheritance, applies to genes in mitochondrial DNA. Because only egg cells contribute mitochondria to the developing embryo, only mothers can pass on mitochondrial conditions to their children. An example of this type of disorder is Leber's hereditary optic neuropathy.

Multifactorial and Polygenic (Complex) Disorders

Genetic disorders may also be complex, multifactorial, or polygenic, meaning that they are likely associated with the effects of multiple genes in combination with lifestyle and environmental factors. Multifactorial disorders include heart disease and diabetes. Although complex disorders often cluster in families, they do not have a clear-cut pattern of inheritance.

This makes it difficult to determine a person's risk of inheriting or passing on these disorders. Complex disorders are also difficult to study and treat because the specific factors that cause most of these disorders have not yet been identified. On a pedigree, polygenic diseases do tend to "run in families", but the inheritance does not fit simple patterns as with Mendelian diseases. But this does not mean that the genes cannot eventually be located and studied. There is also a strong environmental component to many of them (e.g., blood pressure).

- asthma
- autoimmune diseases such as multiple sclerosis
- cancers
- ciliopathies
- cleft palate
- diabetes
- heart disease
- hypertension
- inflammatory bowel disease
- mental retardation
- mood disorder
- obesity
- refractive error
- infertility.

Prognosis and Treatment of Genetic Disorders

Genetic disorders rarely have effective treatments, though gene therapy is being tested as a possible treatment for some genetic diseases, including some forms of retinitis pigmentosa

- Gauchers disease is a genetic disease affecting metabolism. It is more treatable then most other genetic diseases, and can be treated with enzyme replacement therapy, medication miglustat, and bone marrow transplantion.

Huntington's Disease Clinical Research

There are different therapies under investigation for Huntington's disease. Huntington's disease is a genetic neurological disorder characterised after onset by uncoordinated, jerky body movements and a decline in some mental abilities for which there is no cure or effective treatments. Animal models help in the ongoing investigations using intrabody therapy, gene silencing, or stem cell implants, but all are at their first steps. Several clinical trials of various compounds are also in development.

Many of these trials use the unified Huntington's disease rating scale (UHDRS) developed by the Huntington's Study Group. This provides an overall rating system based on motor, behavioural, cognitive, and functional assessments, backed up by a reference database of previous results provided by HSG for reference. Another avenue of research is understanding the physical mechanics of protein folding, HD is one of several targets of the folding@home project, one of the largest distributed computing systems on the World Wide Web.

Animal Models

Appropriate animal models are critical for understanding the fundamental mechanisms causing the disease and for supporting the early stages of drug development. Neurochemically-induced mice or monkeys were first available, but they did not mimic the progressive features of the disease. After the HD gene was discovered, transgenic animals exhibiting HD were generated by inserting a CAG repeat expansion into the genome, of mice (strain R6/2),), *Drosophila* fruit flies, and more recently monkeys. Expression without insertion of a DNA repeat in nematode worms also produced a valuable model.

Intrabody Therapy

Genetically-engineered intracellular antibody fragments called intrabodies have shown therapeutic results, by inhibiting mHtt aggregation, in *Drosophila* models. This was achieved using an intrabody called C4 sFv, a single chain variable fragment which binds to the end of mHtt within a cell. This therapy prevented larval and pupal mortality (without therapy 77% died) and delayed neurodegeneration in the adult, significantly increasing their lifespan. Intrabodies have shown promise as a useful tool for drug discovery, and may potentially be used as a therapy for HD and other neurodegenerative disorders caused by protein misfolding or abnormal protein interactions.

Gene Silencing

As HD has been conclusively linked to a single gene, gene silencing is potentially possible. Researchers have investigated using gene knockdown of mHTT as a potential treatment. Using a mouse model, siRNA therapy achieved a 60% reduction in expression of mHTT and progression of the disease was stalled. However, this study used the human form of the mHTT protein in mice, and was therefore only able to directly target the mHTT, leaving endogenous, wild-type mouse Htt gene expression unaffected.

The widely-held hypothesis that silencing the wild type mouse Htt in middle aged mice, the age of onset of HD, renders undesirable, gene silencing to treat HD, is now largely discredited by experiment. These data indicate that the coincident silencing of the wild-type and mutant htt may be considered as a therapeutic tool for HD. This is despite the unchallenged experimental evidence that wild-type Htt is critical for development in mouse models. Another study showed that mouse models already in late stages of the disease recovered motor function after expression of mHTT was stopped.

Stem Cell Treatments

Stem cell implants are a stem cell treatment where damaged neurons are replaced by transplantation of stem cells (or possibly neural stem cells—a type of somatic [adult] stem cell) into affected regions of the brain. Hypothetically, embryonic stem cells can be differentiated into neuronal precursors *in vitro*, and transplanted into damaged areas of the brain to generate replacement neurons; if enough damaged neurons can be replaced and develop the correct synaptic connectivity, symptoms could be alleviated.

This treatment would not prevent further neuronal damage, so it would have to be an ongoing treatment. Experiments have yielded some positive results in animal models, but stem cell treatment for HD remains highly speculative and has not been tested in clinical trials.

Pharmacological

As of August 2008, several trials of various compounds are in development or ongoing, a few at the point of testing on larger numbers of people, known as phase III of clinical trials. Substances that have shown promise in initial experiments include dopamine receptor blockers, select dopamine antagonists, such as creatine and CoQ10, the antibiotic Minocycline, antioxidant-containing foods and nutrients, and antidepressants, including selective serotonin reuptake inhibitors such as sertraline, fluoxetine, and paroxetine. In 2010, minocycline was found to be ineffective for humans in a multi-centre trial. A 2009 study suggested strong association between specific single nucleotide polymorphisms alleles and CAG expansion also provides an opportunity of personalized therapeutics in HD where the clinical development of only a small number of allele-specific targets may be sufficient to treat up to 88% of the HD patient population.

Other

Recently it was shown that increasing BDNF expression may represent a useful strategy for Huntington's disease treatment.

Research Methods

DNA can be manipulated in the laboratory. Restriction enzymes are commonly used enzymes that cut DNA at specific sequences, producing predictable fragments of DNA. DNA fragments can be visualised through use of gel electrophoresis, which separates fragments according to their length.

The use of ligation enzymes allows DNA fragments to be connected, and by ligating fragments of DNA together from different sources, researchers can create recombinant DNA. Often associated with genetically modified organisms, recombinant DNA is commonly used in the context of plasmids—short circular DNA fragments with a few genes on them. By inserting plasmids into bacteria and growing those bacteria on plates of agar (to isolate clones of bacteria cells), researchers can clonally amplify the inserted fragment of DNA (a process known as molecular cloning). (Cloning can also refer to creating clonal organisms, by various means.)

DNA can also be amplified using a procedure called the polymerase chain reaction (PCR). By using specific short sequences of DNA, PCR can isolate and exponentially amplify a targeted region of DNA. Because it can amplify from extremely small amounts of DNA, PCR is also often used to detect the presence of specific DNA sequences.

DNA Sequencing and Genomics

One of the most fundamental technologies developed to study genetics, DNA sequencing allows researchers to determine the sequence of nucleotides in DNA fragments. Developed in 1977 by Frederick Sanger and coworkers, chain-termination sequencing is now routinely used to sequence DNA fragments. With this technology, researchers have been able to study the molecular sequences associated with many human diseases.

As sequencing has become less expensive, researchers have sequenced the genomes of many organisms, using computational tools to stitch together the sequences of many different fragments (a process called genome assembly). These technologies were used to sequence the human genome, leading to the completion of the Human Genome Project in 2003. New high-throughput sequencing technologies are dramatically lowering the cost of DNA sequencing, with many researchers hoping to bring the cost of resequencing a human genome down to a thousand dollars.

The large amount of sequence data available has created the field of genomics, research that uses computational tools to search for and analyse patterns in the full genomes of organisms. Genomics can also be considered a subfield of bioinformatics, which uses computational approaches to analyse large sets of biological data.

6

Human Stem Cell Lines

Amniotic fluid surrounding the developing fetus contains cells shed by the fetus and is regularly collected from pregnant women during amniocentesis. In 2003, researchers identified a subset of cells in amniotic fluid that express Oct-4, a marker for pluripotent human stem cells that is expressed in ES cells and embryonic germ cells. Since then, investigators have shown that amniotic fluid stem cells can differentiate into cells of all three embryonic germ layers and that these cells do not form tumours in vivo.

For example, Anthony Atala and colleagues at the Wake Forest University have recently generated non-embryonic stem cell lines from cells found in human and rat amniotic fluid. They named these cells amniotic fluid-derived stem cells (AFS). Experiments demonstrate that AFS can produce cells that originate from each of the three embryonic germ layers, and the self-renewing cells maintained the normal number of chromosomes after a prolonged period in culture.

However, undifferentiated AFS did not produce all of the proteins expected of pluripotent cells, and they were not capable of forming a teratoma. The scientists developed in vitro conditions that enabled AFS to produce nerve cells, liver cells, and bone-forming cells. AFS-derived human nerve cells could make proteins typical of specialised nerve cells and were able to integrate into a mouse brain and survive for at least two months. Cultured AFS-derived human liver cells secreted urea and made proteins characteristic of normal human liver cells. Cultured AFS-derived human bone cells made proteins expected of human bone cells and formed bone in mice when seeded onto scaffolds and implanted under the mouse's skin. Although scientists do not yet know how many different cell types AFS can generate, AFS may one day allow researchers to establish a bank of cells for transplantation into humans.

Strategies to "Reprogram" Non-Pluripotent Cells Become Pluripotent Cells

An alternative to searching for an existing population of stem cells is to create a new one from a population of non-pluripotent cells. This strategy, which may or may not involve the creation of an embryo, is known as "reprogramming." This section will summarize reprogramming approaches, including several recent breakthroughs in the field..

Reprogramming Through Somatic Cell Nuclear Transfer (SCNT)

In SCNT, human oocytes (eggs) are collected from a volunteer donor who has taken drugs that stimulate the production of more than one oocyte during the menstrual cycle. Scientists then remove the nucleus from the donated oocyte and replace it with the nucleus from a somatic cell, a differentiated adult cell from elsewhere in the body. The oocyte with the newly-transferred nucleus is then stimulated to develop. The oocyte may develop only if the transplanted nucleus is returned to the pluripotent state by factors present in the oocyte cytoplasm. This alteration in the state of the mature nucleus is called nuclear reprogramming. When development progresses to the blastocyst stage, the ICM is removed and placed into culture in an attempt to establish a pluripotent stem cell line. To date, the technique has been successfully demonstrated in two primates: macaque monkeys and humans.

However, successful SCNT creates an embryo-like entity, thereby raising the ethical issues that confront the use of spare IVF embryos. However, pluripotent cell lines created by embryos generated by SCNT offer several advantages over ES cells. First, the nuclear genes of such a pluripotent cell line will be identical to the genes in the donor nucleus. If the nucleus comes from a cell that carries a mutation underlying a human genetic disease such as Huntington's disease, then all cells derived from the pluripotent cell line will carry this mutation.

In this case, the SCNT procedure would enable the development of cellular models of human genetic disease that can inform our understanding of the biology of disease and facilitate development of drugs to slow or halt disease progression. Alternatively, if the cell providing the donor nucleus comes from a specific patient, all cells derived from the resulting pluripotent cell line will be genetically matched to the patient with respect to the nuclear genome. If these cells were used in transplantation therapy, the likelihood that the patient's immune system would recognize the transplanted cells as foreign and initiate tissue rejection would be reduced. However, because mitochondria also contain DNA, the donor oocyte will be the source of the mitochondrial genome, which is likely to carry mitochondrial gene differences from the patient which may still lead to tissue rejection.

A technique reported in 2007 by Dr. Kevin Eggan and colleagues at Harvard University may expand scientists' options when trying to "reprogram" an adult cell's DNA. Previously, successful SCNT relied upon the use of an unfertilized egg. Now, the Harvard scientists have demonstrated that by using a drug to stop cell division in a fertilized mouse egg (zygote) during mitosis, they can successfully reprogram an adult mouse skin cell by taking advantage of the "reprogramming factors" that are active in the zygote at mitosis. They removed the chromosomes from the single-celled zygote's nucleus and replaced them with the adult donor cell's chromosomes. The active reprogramming factors present in the zygote turned genes on and off in the adult donor chromosomes, to make them behave like the chromosomes of a normally fertilized zygote. After the zygote was stimulated to divide, the cloned mouse embryo developed to the blastocyst stage, and the scientists were able to harvest embryonic stem cells from the resulting blastocyst. When the scientists applied their new method to abnormal mouse zygotes, they succeeded at reprogramming adult mouse skin cells and harvesting stem cells. If this technique can be repeated with abnormal human zygotes created in excess after IVF procedures, scientists could use them for research instead of discarding them as medical waste.

Reprogramming through Altered Nuclear Transfer (ANT)

Altered nuclear transfer is a variation on standard SCNT that proposes to create patient-specific stem cells without destroying an embryo. In ANT, scientists turn off a gene needed for implantation in the uterus (Cdx2) in the patient cell nucleus before it is transferred into the donor egg. In 2006, Dr. Rudolph Jaenisch and colleagues at MIT demonstrated that ANT can be carried out in mice. Mouse ANT entities whose Cdx2 gene is switched off are unable to implant in the uterus and do not survive to birth. Although ANT has been used to create viable stem cell lines capable of producing almost all cell types, the authors point out that this technique must still be tested with monkey and human embryos. Moreover, the manipulation needed to control Cdx2 expression introduces another logistical hurdle that may complicate the use of ANT to derive embryonic stem cells. Proponents of ANT, such as William Hurlbut of the Stanford University Medical Centre, suggest that the entity created by ANT is not a true embryo because it cannot implant in the uterus. However, the technique is highly controversial, and its ethical implications remain a source of current debate.

Reprogramming through Cell Fusion

In 2005, Kevin Eggan and colleagues at Harvard University reported that they had fused cultured adult human skin cells with hESCs. The resulting "hybrid" cells featured many characteristics of hESCs, including a similar manner of growth and division and the manufacture of proteins typically produced by hESCs. Some factor(s) within the hESCs enabled them to "reprogram" the adult skin cells to behave as hESCs. However, these cells raised a significant technical barrier to clinical use. Because fused cells are tetraploid (they contain four copies of the cellular DNA rather than the normal two copies), scientists would need to develop a method to remove the extra DNA without eliminating their hESC-like properties. The fusion method serves as a useful model system for studying how stem cells "reprogram" adult cells to have properties of pluripotent cells. However, if the reprogramming technique could be carried out without the fusion strategy, a powerful avenue for creating patient-specific stem cells without using human eggs could be developed.

Induced Pluripotent Stem Cells (iPSCs): Reprogramming Adult Somatic Cells to Become Pluripotent Stem Cells

In 2007, two independent research groups published manuscripts that described successful genetic reprogramming of human adult somatic cells into pluripotent human stem cells. Although some technical limitations remain, this strategy suggests a promising new avenue for generating pluripotent cell lines that can inform drug development, models of disease, and ultimately, transplantation medicine. These experiments, which are discussed below, were breakthroughs because they used adult somatic cells to create pluripotent stem cells that featured hallmarks of ES cells.

In 2006, Shinya Yamanaka and colleagues at Kyoto University reported that they could use a retroviral expression vector to introduce four important stem cell factors into adult mouse cells and reprogram them to behave like ES cells. They called the reprogrammed cells "iPSCs," for induced pluripotent stem cells. However, iPSCs produced using the original technique failed to produce sperm and egg cells when injected into an early mouse blastocyst and did not make certain critical DNA changes. These researchers then modified the technique to select for iPSCs that can produce sperm and eggs, results that have since been reproduced by Rudolph Jaenisch and colleagues at the Massachusetts Institute of Technology (MIT).

In addition, the MIT scientists determined that iPSCs DNA is modified in a manner similar to ES cells, and important stem cell genes are expressed at similar levels. They also demonstrated that iPSCs injected into an early mouse blastocyst can produce all cell types within the developing embryo, and such embryos can complete gestation and are born alive.

Once these research advances were made in mice, they suggested that similar techniques might be used to reprogram adult human cells. In 2007, Yamanaka and coworkers reported that introducing the same four genetic factors that reprogrammed the mouse cells into adult human dermal fibroblasts reprogrammed the cells into human iPSCs. These iPSCs were similar to human ES cells in numerous ways, including morphology, proliferative capacity, expression of cell surface antigens, and gene expression. Moreover, the cells could differentiate into cell types from the three embryonic germ layers both in vitro and in teratoma assays. Concurrent with the Yamanaka report, James Thomson and coworkers at the University of Wisconsin published a separate manuscript that detailed the creation of human iPSCs through somatic cell reprogramming using four genetic factors (two of which were in common with the Yamanaka report). The cells generated by the Thomson group met all defining criteria for ES cells, with the exception that they were not derived from embryos.

These breakthroughs have spurred interest in the field of iPSCs research. In early 2008, investigators at the Massachusetts General Hospital and the University of California, Los Angeles reported generating reprogrammed cells. As scientists explore the mechanisms that govern reprogramming, it is anticipated that more reports will be forthcoming in this emerging area. Although these reprogramming methods require the use of a virus, non-viral strategies may also be possible in the future. In any case, these approaches have created powerful new tools to enable the "de-differentation" of cells that scientists had previously believed to be terminally differentiated.

Although further study is warranted to determine if iPS and ES cells differ in clinically significant ways, these breakthrough reports suggest that reprogramming is a promising strategy for future clinical applications. Induced pluripotent cells offer the obvious advantage that they are not derived from embryonic tissues, thereby circumventing the ethical issues that surround use of these materials. Successful reprogramming of adult somatic cells could also lead to the development of stem cell lines from patients who suffer from genetically-based diseases, such as Huntington's Disease, spinal muscular atrophy, muscular dystrophy, and thalessemia. These lines would be invaluable research tools to understand the mechanisms of these diseases and to test potential drug treatments. Additionally, reprogrammed cells could potentially be used to repair damaged tissues; patient-specific cell lines could greatly reduce the concerns of immune rejection that are prevalent with many transplantation strategies.

However, several technical hurdles must be overcome before iPSCs can be used in humans. For example, in preliminary experiments with mice, the virus used to introduce the stem cell factors sometimes caused cancers. The viral vectors used in these experiments will have to be selected carefully and tested fully to verify that they do not integrate into the genome, thereby harbouring the potential to introduce genetic mutations at their site of insertion. This represents a significant concern that must be addressed before the technique can lead to useful treatments for humans. However, this strategy identifies a method for creating pluripotent stem cells that, together with studies of other types of pluripotent stem cells, will help researchers learn how to reprogram cells to repair damaged tissues in the human body.

Other Sources of Pluripotent and/or Multipotent Cells

Stem cell research is a rapidly evolving field, and researchers continue to isolate new pluripotent cells and create additional cell lines. This section briefly reviews other sources of pluripotent cells and the implications that their discovery may have on future research. Epiblast Cells. While rodent and human ES cells are pluripotent, they maintain their respective pluripotencies through different molecular signalling pathways. It is not known why these differences exist. Recently, several research groups have reported the generation of stable, pluripotent cell lines from mouse and rat epiblast, a tissue of the post-implantation embryo that ultimately generates the embryo proper.

These cells are distinct from mouse ES cells in terms of the signals that control their differentiation. However, the cells share patterns of gene expression and signalling responses with human ES cells. The establishment of epiblast cell lines can therefore provide insight into the distinctions between pluripotent cells from different species and illuminate ways that pluripotent cells pursue distinct fates during early development. Existing Adult Stem Cells. As has been discussed in other chapters, numerous types of precursor cells have been isolated in adult tissues. Although these cells tend to be relatively rare and are dispersed throughout the tissues, they hold great potential for clinical application and tissue engineering. For example, tissues created using stem cells harvested from an adult patient could theoretically be used clinically in that patient without engendering an immune response.

Moreover, the use of adult stem cells avoids the ethical concerns associated with the use of ES cells. In addition, adult-derived stem cells do not spontaneously differentiate as do ES cells, thus eliminating the formation of teratomas often seen with implantation of ES cells. The potential of adult stem cells for regenerative medicine is great; it is likely that these various cells will find clinical application in the upcoming decades.

Pluripotent Cell Lines are Tools for Future Research

Although the recent advances in reprogramming of adult somatic cells has generated a wave of interest in the scientific community, these cell lines will not likely replace hESC lines as tools for research and discovery. Rather, both categories of cells will find unique uses in the study of stem cell biology and the development and evaluation of therapeutic strategies. Pluripotent cells offer a number of potential clinical applications, especially for diseases with a genetic basis. However, researchers are just beginning to unlock the many factors that govern the cells' growth and differentiation. As scientists make strides toward understanding how these cells can be manipulated, additional applications, approaches, and techniques will likely emerge. As such, pluripotent cells will play a pivotal role in future research into the biology of development and the treatment of disease.

Bone Marrow

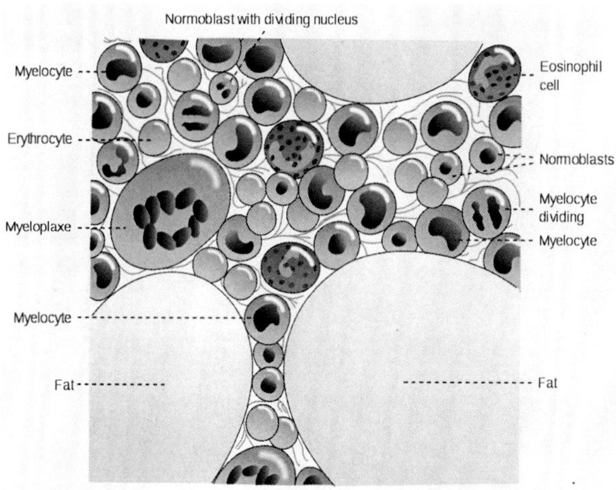

Figure: *Simplified illustration of cells in bone marrow.*

Bone marrow is the flexible tissue found in the interior of bones. In humans, bone marrow in large bones produces new blood cells. On average, bone marrow constitutes 4% of the total body mass of humans; in adults weighing 65 kg (143 lbs), bone marrow accounts for approximately 2.6 kg (5.7 lbs). The hematopoietic compartment of bone marrow produces approximately 500 billion blood cells per day, which use the bone marrow vasculature as a conduit to the body's systemic circulation.

Marrow Types

There are two types of bone marrow: medulla ossium rubra (*red marrow* - consisting mainly of hematopoietic tissue) and medulla ossium flava (*yellow marrow* - consisting mainly of fat cells). Red blood cells, platelets and most white blood cells arise in red marrow. Both types of bone marrow contain numerous blood vessels and capillaries. At birth, all bone marrow is red. With age, more and more of it is converted to the yellow type; only around half of adult bone marrow is red. Red marrow is found mainly in the flat bones, such as the hip bone, breast bone, cranium, ribs, vertebrae and shoulder blades, and in the cancellous ("spongy") material at the epiphyseal ends of long bones such as the femur and humerus.

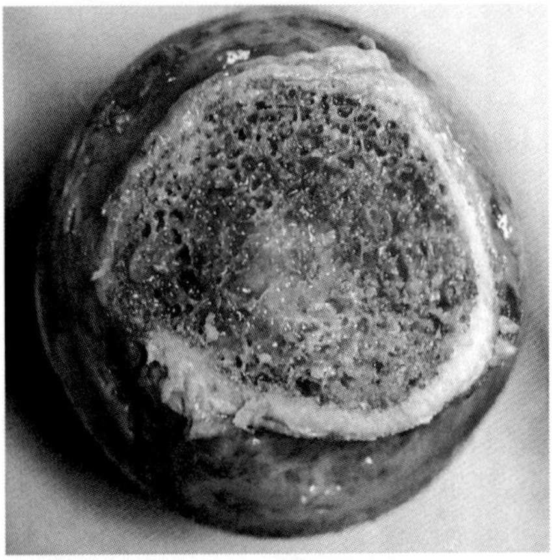

Figure: *A femur with a cortex of cortical bone and medulla of trabecular bone, showing its red bone marrow and a focus of yellow bone marrow.*

Yellow marrow is found in the medullary cavity, the hollow interior of the middle portion of long bones. In cases of severe blood loss, the body can convert yellow marrow back to red marrow to increase blood cell production.

Stroma

The stroma of the bone marrow is all tissue not directly involved in the primary function of hematopoiesis. Yellow bone marrow makes up the majority of bone marrow stroma, in addition to smaller concentrations of stromal cells located in the red bone marrow.

Though not as active as parenchymal red marrow, stroma is indirectly involved in hematopoiesis, since it provides the hematopoietic microenvironment that facilitates hematopoiesis by the parenchymal cells. For instance, they generate colony stimulating factors, which have a significant effect on hematopoiesis. Cells that constitute the bone marrow stroma are:

- fibroblasts (reticular connective tissue)
- macrophages
- adipocytes
- osteoblasts
- osteoclasts
- endothelial cells, which form the sinusoids. These derive from endothelial stem cells, which are also present in the bone marrow.

Macrophages contribute especially to red blood cell production, as they deliver iron for hemoglobin production.

Bone Marrow Barrier

The blood vessels of the bone marrow constitute a barrier, inhibiting immature blood cells from leaving the marrow. Only mature blood cells contain the membrane proteins required to attach to and pass the blood vessel endothelium. Hematopoietic stem cells may also cross the bone marrow barrier, and may thus be harvested from blood.

Mesenchymal Stem Cells

The bone marrow stroma contain mesenchymal stem cells (MSCs), also called *marrow stromal cells*. These are multipotent stem cells that can differentiate into a variety of cell types. MSCs have been shown to differentiate, in vitro or in vivo, into osteoblasts, chondrocytes, myocytes, adipocytes and beta-pancreatic islets cells. MSCs can also transdifferentiate into neuronal cells.

Red Marrow Parenchyma

Types of Cells

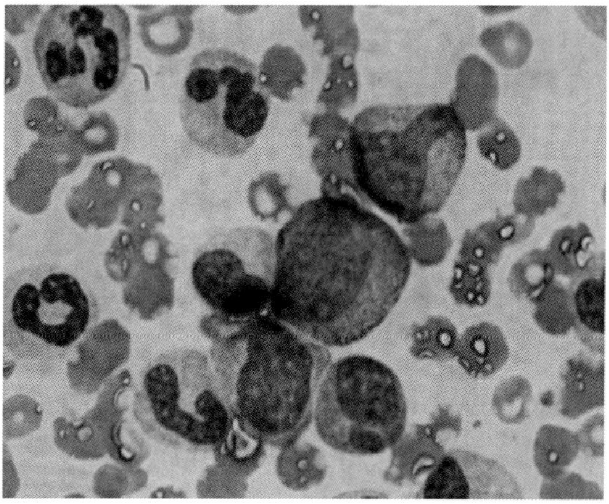

Figure: *Hematopoietic precursor cells: promyelocyte in the centre, two metamyelocytes next to it and band cells from a bone marrow aspirate.*

Table: *Cellular constitution of the red bone marrow parenchyma*

Group	*Cell type*	*Average fraction*
Cells of myelopoiesis	Myeloblasts	0.9%
	Promyelocytes	3.3%
	Neutrophilic myelocytes	12.7%

	Eosinophilic myelocytes	0.8%
	Neutrophilic metamyelocytes	15.9%
	Eosinophilic metamyelocytes	1.2%
	Neutrophilic band cells	12.4%
	Eosinophilic band cells	0.9%
	Segmented neutrophils	7.4%
	Segmented eosinophils	0.5%
	Segmented basophils +	0.1%
Cells of erythropoiesis	mast cells Pronormoblasts	0.6%
	Basophilic normoblasts	1.4%
	Polychromatic normoblasts	21.6%
	Orthochromatic normoblast	2.0%
Other cell types	Megakaryocytes	< 0.1%
	Plasma cells	1.3%
	Reticular cells	0.3%
	Lymphocytes	16.2%
	Monocytes	0.3%

In addition, the bone marrow contains hematopoietic stem cells, which give rise to the three classes of blood cells that are found in the circulation: white blood cells (leukocytes), red blood cells (erythrocytes), and platelets (thrombocytes).

Compartmentalisation

There is biologic compartmentalisation in the bone marrow, in that certain cell types tend to aggregate in specific areas. For instance, erythrocytes, macrophages, and their precursors tend to gather around blood vessels, while granulocytes gather at the borders of the bone marrow.

Lymphatic Role

The red bone marrow is a key element of the lymphatic system, being one of the primary lymphoid organs that generate lymphocytes from immature hematopoietic progenitor cells. The bone marrow and thymus constitute the primary lymphoid tissues involved in the production and early selection of lymphocytes. Furthermore, bone marrow performs a valve-like function to prevent the backflow of lymphatic fluid in the lymphatic system.

Diseases Involving the Bone Marrow

The normal bone marrow architecture can be displaced by malignancies, aplastic anemia, or infections such as tuberculosis, leading to a decrease in the production of blood cells and blood platelets. In addition, cancers of the hematologic progenitor cells in the bone marrow can arise; these are the leukemias.

Exposure to radiation or chemotherapy will kill many of the rapidly dividing cells of the bone marrow, and will therefore result in a depressed immune system. Many of the symptoms of radiation sickness are due to damage to the bone marrow cells.

To diagnose diseases involving the bone marrow, a bone marrow aspiration is sometimes performed. This typically involves using a hollow needle to acquire a sample of red bone marrow from the crest of the ilium under general or local anesthesia.

Examination

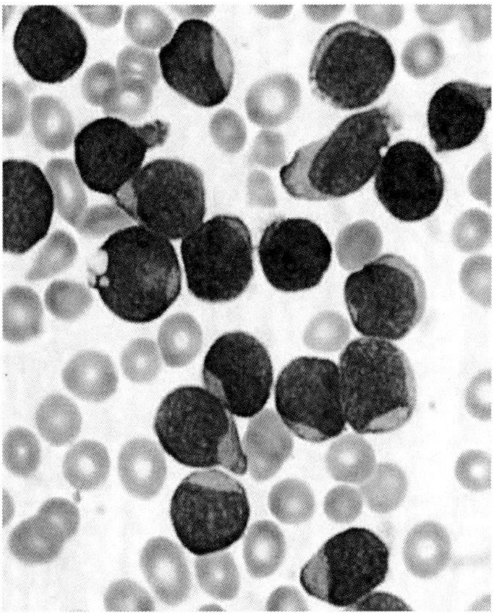

Figure: *A Wright's-stained bone marrow aspirate smear from a patient with leukemia.*

Bone marrow examination is the pathologic analysis of samples of bone marrow obtained by *bone marrow biopsy* and *bone marrow aspiration*. Bone marrow examination is used in the diagnosis of a number of conditions, including leukemia, multiple myeloma, anemia, and pancytopenia.

The bone marrow produces the cellular elements of the blood, including platelets, red blood cells and white blood cells.

While much information can be gleaned by testing the blood itself (drawn from a vein by phlebotomy), it is sometimes necessary to examine the source of the blood cells in the bone marrow to obtain more information on hematopoiesis; this is the role of bone marrow aspiration and biopsy.

Donation and Transplantation of Bone Marrow

In a bone marrow transplant, hematopoietic stem cells are removed from a person and infused into another person (allogenic) or into the same person at a later time (autologous). If donor and recipient are compatible, these infused cells will then travel to the bone marrow and initiate blood cell production. Transplantation from one person to another is performed in severe cases of disease of the bone marrow. The patient's marrow is first killed off with drugs or radiation, and then the new stem cells are introduced. Before radiation therapy or chemotherapy in cases of cancer, some of the patient's hematopoietic stem cells are sometimes harvested and later infused back when the therapy is finished to restore the immune system.

Harvesting

The stem cells are harvested directly from the red marrow in the crest of the ilium, often under general anesthesia. The procedure is minimally invasive and does not require stitches afterwards. Depending on the donor health and reaction to the procedure, the actual harvesting can be an outpatient procedure or requiring 1–2 days of recovery in the hospital. Another option is to administer certain drugs that stimulate the release of stem cells from the bone marrow into circulating blood. An IV is inserted into the donor's arm, and the stem cells are filtered out of the blood. The procedure is similar to donating blood or platelets. It may also be taken from the sternum.

The tibia may seem a good source, since it is very superficial. However, except in children, this bone marrow does not contain any substantial amount of red bone marrow, only yellow bone marrow. In newborns, stem cells may be retrieved from the umbilical cord.

Bone Marrow as a Food

Many cultures have utilised bone marrow as a food throughout history. Some anthropologists believe that early humans were scavengers rather than hunters. Marrow would then have been a food source (largely for its fat content) for tool-using hominids, who were able to crack open the bones of carcasses left by apex predators such as lions. In Vietnam, beef bone marrow is used as the soup base for the national staple dish, *phß*, while in the Philippines, the soup *bulalo* is made primarily of beef stock and marrow bones, seasoned with vegetables and boiled meat; a similar soup in the Philippines is called *kansi*.

In Indonesia, bone marrow is called *sumsum* and can be found especially in Minangkabau cuisine. *Sumsum* is often cooked as soup or as *gulai* (a curry-like dish). In India and Pakistan, slow-cooked marrow is the core ingredient of the dish *nalli nihari*. In Hungary, tibia is a main ingredient of beef soup;

the bone is chopped into short (10–15 cm) pieces and the ends are covered with salt to prevent the marrow from leaking from the bone while cooking. Upon serving the soup, the marrow is usually spread on toast.

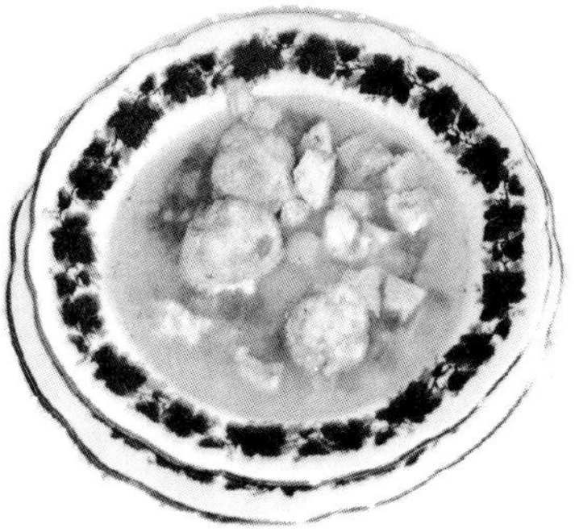

Figure: *In some parts of Germany, beef soup is served with "Markklößchen" (bone marrow balls).*

Beef bone marrow is also the main ingredient in Italian dish ossobuco (braised veal shanks), and beef marrow bones are often included in the French pot-au-feu broth, the cooked marrow being traditionally eaten on toasted bread with sprinkled coarse sea salt. In Iranian cuisine, lamb shanks are usually broken before cooking to allow diners to suck out and eat the marrow when the dish is served. Similar practices are also common in Pakistani cuisine. Some Native Alaskans eat the bone marrow of caribou and moose. European diners in the 18th century often used a marrow scoop (or marrow spoon), often of silver and with a long thin bowl, as a table implement for removing marrow from a bone.

Though once used in various preparations, including pemmican, bone marrow's popularity is now relatively limited in the western world. Nonetheless, it remains in use in some gourmet restaurants and is popular among food enthusiasts.

Stem Cell Treatments and Veterinary Applications

Stem cell treatments are a type of intervention strategy that introduces new cells into damaged tissue in order to treat disease or injury. Many medical researchers believe that stem cell treatments have the potential to change the face of human disease and alleviate suffering. The ability of stem cells to self-renew and give rise to subsequent generations with variable degrees of differentiation capacities, offers significant potential for generation of tissues that can potentially replace diseased and damaged areas in the body, with minimal risk of rejection and side effects.

A number of stem cell therapies exist, but most are at experimental stages or costly, with the notable exception of bone marrow transplantation. Medical researchers anticipate that adult and embryonic stem cells will soon be able to treat cancer, Type 1 diabetes mellitus, Parkinson's disease, Huntington's disease, Celiac Disease, cardiac failure, muscle damage and neurological disorders, and many others. Nevertheless, before stem cell therapeutics can be applied in the clinical setting, more research is necessary to understand stem cell behaviour upon transplantation as well as the mechanisms of stem cell interaction with the diseased/injured microenvironment.

Current Treatments

For over 30 years, bone marrow, and more recently, umbilical cord blood stem cells, have been used to treat cancer patients with conditions such as leukemia and lymphoma. During chemotherapy, most growing cells are killed by the cytotoxic agents. These agents, however, cannot discriminate between the leukemia or neoplastic cells, and the hematopoietic stem cells within the bone marrow. It is this side effect of conventional chemotherapy strategies that the stem cell transplant attempts to reverse; a donor's healthy bone marrow reintroduces functional stem cells to replace the cells lost in the host's body during treatment.

Potential Treatments

Brain Damage

Stroke and traumatic brain injury lead to cell death, characterised by a loss of neurons and oligodendrocytes within the brain. Healthy adult brains contain neural stem cells which divide to maintain general stem cell numbers, or become progenitor cells. In healthy adult animals, progenitor cells migrate within the brain and function primarily to maintain neuron populations for olfaction (the sense of smell). Interestingly, in pregnancy and after injury, this system appears to be regulated by growth factors and can increase the rate at which new brain matter is formed. Although the reparative process appears to initiate following trauma to the brain, substantial recovery is rarely observed in adults, suggesting a lack of robustness. Stem cells may also be used to treat brain degeneration, such as in Parkinson's and Alzheimer's disease.

Cancer

The development of gene therapy strategies for treatment of intracranial tumours offers much promise, and has shown to be successful in the treatment of some dogs; although research in this area is still at an early stage. Using conventional techniques, brain cancer is difficult to treat because it spreads so rapidly. Researchers at the Harvard Medical School transplanted human neural stem cells into the brain of rodents that received intracranial tumours. Within days, the cells migrated into the cancerous area and produced cytosine deaminase, an enzyme that converts a nontoxic prodrug into a chemotheraputic agent. As a result, the injected substance was able to reduce the tumor mass by 81 percent. The stem cells neither differentiated nor turned tumorigenic. Some researchers believe that the key to finding a cure for cancer is to inhibit proliferation of cancer stem cells. Accordingly, current cancer treatments are designed to kill cancer cells. However, conventional chemotherapy treatments cannot discriminate between cancerous cells and others. Stem cell therapies may serve as potential treatments for cancer. Research on treating Lymphoma using adult stem cells is underway and has had human trials. Essentially, chemotherapy is used to completely destroy the patients own lymphocytes, and stem cells injected, eventually replacing the immune system of the patient with that of the healthy donor.

Spinal Cord Injury

A team of Korean researchers reported on November 25, 2003, that they had transplanted multipotent adult stem cells from umbilical cord blood to a patient suffering from a spinal cord injury and that following the procedure, she could walk on her own, without difficulty. The patient had not been able to stand up for roughly 19 years. For the unprecedented clinical test, the scientists isolated adult stem cells from umbilical cord blood and then injected them into the damaged part of the spinal cord.

According to the October 7, 2005 issue of *The Week*, University of California, Irvine researchers transplanted multipotent human fetal-derived neural stem cells into paralysed mice, resulting in locomotor improvements four months later. The observed recovery was associated with differentiation of transplanted cells into new neurons and oligodendrocytes- the latter of which forms the myelin sheath around axons of the central nervous system, thus insulating neural impulses and facilitating communication with the brain.

In January 2005, researchers at the University of Wisconsin–Madison differentiated human blastocyst stem cells into neural stem cells, then into premature motor neurons, and finally into spinal motor neurons, the cell type that, in the human body, transmits messages from the brain to the spinal cord and subsequently mediates motor function in the periphery. The newly generated motor neurons exhibited electrical activity, the signature action of neurons. Lead researcher Su-Chun Zhang described the process as "[teaching] the blastocyst stem cells to change step by step, where each step has different conditions and a strict window of time."

Transformation of blastocyst stem cells into motor neurons had eluded researchers for decades. While Zhang's findings were a significant contribution to the field, the ability of transplanted neural cells to establish communication with neighbouring cells remains unclear. Accordingly, studies using chicken embryos as a model organism can be an effective proof-of-concept experiment.

If functional, the new cells could be used to treat diseases like Lou Gehrig's disease, muscular dystrophy, and spinal cord injuries.

Heart Damage

Several clinical trials targeting heart disease have shown that adult stem cell therapy is safe, effective, and equally efficient in treating old and recent infarcts. Adult stem cell therapy for treating heart disease was commercially available in at least five continents at the last count (2007).

Possible mechanisms of recovery include:

- Generation of heart muscle cells
- Stimulation of growth of new blood vessels to repopulate damaged heart tissue
- Secretion of growth factors
- Assistance via some other mechanism.

It may be possible to have adult bone marrow cells differentiate into heart muscle cells.

Haematopoiesis (Blood Cell Formation)

The specificity of the human immune cell repertoire is what allows the human body to defend itself from rapidly adapting antigens. However, the immune system is vulnerable to degradation upon the pathogenesis of disease, and because of the critical role that it plays in overall defence, its degradation is often fatal to the organism as a whole. Diseases of hematopoietic cells are called hematopathology.

The specificity of the immune cells is what allows recognition of foreign antigens, causing further challenges in the treatment of immune disease. Identical matches between donor and recipient must be made for successful transplantation treatments, but matches are uncommon, even between first-degree relatives. Research using both hematopoietic adult stem cells and embryonic stem cells has provided insight into the possible mechanisms and methods of treatment for many of these ailments.

Fully mature human red blood cells may be generated *ex vivo* by hematopoietic stem cells (HSCs), which are precursors of red blood cells. In this process, HSCs are grown together with stromal cells, creating an environment that mimics the conditions of bone marrow, the natural site of red blood cell growth. Erythropoietin, a growth factor, is added, coaxing the stem cells to complete terminal differentiation into red blood cells.

Further research into this technique should have potential benefits to gene therapy, blood transfusion, and topical medicine.

Baldness

Hair follicles also contain stem cells, and some researchers predict research on these follicle stem cells may lead to successes in treating baldness through an activation of the stem cells progenitor cells. This treatment is expected to work by activating already existing stem cells on the scalp. Later treatments may be able to simply signal follicle stem cells to give off chemical signals to nearby follicle cells which have shrunk during the aging process, which in turn respond to these signals by regenerating and once again making healthy hair. Most recently, Dr. Aeron Potter of the University of California has claimed that stem cell therapy led to a significant and visible improvement in follicular hair growth. Results from his experiments are under review by the journal Science (journal).

Missing Teeth

In 2004, scientists at King's College London discovered a way to cultivate a complete tooth in mice and were able to grow them stand-alone in the laboratory. Researchers are confident that this technology can be used to grow live teeth in human patients.

In theory, stem cells taken from the patient could be coaxed in the lab into turning into a tooth bud which, when implanted in the gums, will give rise to a new tooth, and would be expected to grow within two months. It will fuse with the jawbone and release chemicals that encourage nerves and blood vessels to connect with it. The process is similar to what happens when humans grow their original adult teeth. Many challenges remain, however, before stem cells could be a choice for the replacement of missing teeth in the future.

Deafness

Heller has reported success in regrowing cochlea hair cells with the use of embryonic stem cells.

Blindness and Vision Impairment

Since 2003, researchers have successfully transplanted corneal stem cells into damaged eyes to restore vision. "Sheets of retinal cells used by the team are harvested from aborted fetuses, which some people find objectionable." When these sheets are transplanted over the damaged cornea, the stem cells stimulate renewed repair, eventually restore vision. The latest such development was in June 2005, when researchers at the Queen Victoria Hospital of Sussex, England were able to restore the sight of forty patients using the same technique. The group, led by Dr. Sheraz Daya, was able to successfully use adult stem cells obtained from the patient, a relative, or even a cadaver. Further rounds of trials are ongoing.

In April 2005, doctors in the UK transplanted corneal stem cells from an organ donor to the cornea of Deborah Catlyn, a woman who was blinded in one eye when acid was thrown in her eye at a nightclub. The cornea, which is the transparent window of the eye, is a particularly suitable site for transplants. In fact, the first successful human transplant was a cornea transplant. The absence of blood vessels within the cornea makes this area a relatively easy target for transplantation. The majority of corneal transplants carried out today are due to a degenerative disease called keratoconus.

The University Hospital of New Jersey reports that the success rate for growth of new cells from transplanted stem cells varies from 25 percent to 70 percent. In 2009, researchers at the University of Pittsburgh Medical centre demonstrated that stem cells collected from human corneas can restore transparency without provoking a rejection response in mice with corneal damage.

Amyotrophic Lateral Sclerosis

Stem cells have resulted in significant locomotor improvements in rats with an Amyotrophic lateral sclerosis-like disease. In a rodent model that closely mimics the human form of ALS, animals were injected with a virus to kill the spinal cord motor nerves which mediate movement. Animals subsequently received stem cells in the spinal cord. Transplanted cells migrated to the sites of injury, contributed to regeneration of the ablated nerve cells, and restored locomotor function.

Graft Vs. Host Disease and Crohn's Disease

Phase III clinical trials expected to end in second-quarter 2008 were conducted by Osiris Therapeutics using their in-development product Prochymal, derived from adult bone marrow. The target disorders of this therapeutic are graft-versus-host disease and Crohn's disease.

Neural and Behavioural Birth Defects

A team of researchers led by Prof. Joseph Yanai were able to reverse learning deficits in the offspring of pregnant mice who were exposed to heroin and the pesticide organophosphate. This was done by direct neural stem cell transplantation into the brains of the offspring. The recovery was almost 100 percent, as shown both in behavioural tests and objective brain chemistry tests. Behavioural tests and learning scores of the treated mice showed rapid improvement after treatment, providing results that rivalled non-treated mice. On the molecular level, brain chemistry of the treated animals was also restored to normal. Through the work, which was supported by the US National Institutes of Health, the US-Israel Binational Science Foundation and the Israel anti-drug authorities, the researchers discovered that the stem cells worked even in cases where most of the cells died out in the host brain. The scientists found that before they die the neural stem cells succeed in inducing the host brain to produce large numbers of stem cells which repair the damage. These findings, which answered a major question in the stem cell research community, were published in January 2008 in the leading journal, Molecular Psychiatry.

Scientists are now developing procedures to administer the neural stem cells in the least invasive way possible - probably via blood vessels, making therapy practical and clinically feasible. Researchers also plan to work on developing methods to take cells from the patient's own body, turn them into stem cells, and then transplant them back into the patient's blood via the blood stream. Aside from decreasing the chances of immunological rejection, the approach will also eliminate the controversial ethical issues involved in the use of stem cells from human embryos.

Diabetes

Diabetes patients lose the function of insulin-producing beta cells within the pancreas. Human embryonic stem cells may be grown in cell culture and stimulated to form insulin-producing cells that can be transplanted into the patient. However, clinical success is highly dependent on the development of the following procedures:

- Transplanted cells should proliferate
- Transplanted cells should differentiate in a site-specific manner
- Transplanted cells should survive in the recipient (prevention of transplant rejection)
- Transplanted cells should integrate within the targeted tissue
- Transplanted cells should integrate into the host circuitry and restore function.

Orthopaedics

Clinical case reports in the treatment of orthopaedic conditions have been reported. To date, the focus in the literature for musculoskeletal care appears to be on mesenchymal stem cells. Centeno et al. have published MRI evidence of increased cartilage and meniscus volume in individual human subjects. The results of trials that include a large number of subjects, are yet to be published. However, a published safety study conducted in a group of 227 patients over a 3-4 year period shows adequate safety and minimal complications associated with mesenchymal cell transplantation. Wakitani has also published a small case series of nine defects in five knees involving surgical transplantation of mesenchymal stem cells with coverage of the treated chondral defects.

Wound Healing

Stem cells can also be used to stimulate the growth of human tissues. In an adult, wounded tissue is most often replaced by scar tissue, which is characterised in the skin by disorganised collagen structure, loss of hair follicles and irregular vascular structure. In the case of wounded fetal tissue, however, wounded tissue is replaced with normal tissue through the activity of stem cells.

A possible method for tissue regeneration in adults is to place adult stem cell "seeds" inside a tissue bed "soil" in a wound bed and allow the stem cells to stimulate differentiation in the tissue bed cells. This method elicits a regenerative response more similar to fetal wound-healing than adult scar tissue formation. Researchers are still investigating different aspects of the "soil" tissue that are conducive to regeneration.

Infertility

Culture of human embryonic stem cells in mitotically inactivated porcine ovarian fibroblasts (POF) causes differentiation into germ cells (precursor cells of oocytes and spermatozoa), as evidenced by gene expression analysis.

Human embryonic stem cells have been stimulated to form Spermatozoon-like cells, yet still slightly damaged or malformed. It could potentially treat azoospermia.

Clinical Trials

On January 23, 2009, the US Food and Drug Administration gave clearance to Geron Corporation for the initiation of the first clinical trial of an embryonic stem cell-based therapy on humans. The trial will evaluate the drug GRNOPC1, embryonic stem cell-derived oligodendrocyte progenitor cells, on patients with acute spinal cord injury. As of mid 2010 hundreds of phase III clinical trials involving stem cells have been registered.

Potential Contributions to Veterinary Medicine

Research currently conducted on horses, dogs, and cats can benefit the development of stem-cell treatments in veterinary medicine and can target a wide range of injuries and diseases such as myocardial infarction, stroke, tendon and ligament damage, osteoarthritis, osteochondrosis and muscular dystrophy both in large animals, as well as humans.

While investigation of cell-based therapeutics generally reflects human medical needs, the high degree of frequency and severity of certain injuries in racehorses has put veterinary medicine at the forefront of this novel regenerative approach. Companion animals can serve as clinically relevant models that closely mimic human disease.

Development of Regenerative Treatment Models

Veterinary applications of stem cell therapy as a means of tissue regeneration have been largely shaped by research that began with the use of adult-derived mesenchymal stem cells to treat animals with injuries or defects affecting bone, cartilage, ligaments and/or tendons. Because mesenchymal stem cells can differentiate into the cells that make up bone, cartilage, tendons, and ligaments (as well as muscle, fat, and possibly other tissues), they have been the main type of stem cells studied in the treatment of diseases affecting these tissues. Mesenchymal stem cells are primarily derived from adipose tissue or bone marrow.

Since an elevated immune response following cell transplantation may result in rejection of exogenous cells (except in the case of cells derived from a very closely genetically related individual), mesenchymal stem cells are often derived from the patient prior to injection in a process known as autologous transplantation. Surgical repair of bone fractures in dogs and sheep has demonstrated that engraftment of mesenchymal stem cells derived from a genetically different donor within the same species, termed allogeneic transplantation, does not elicit an immunological response in the recipient animal and can mediate regeneration of bone tissue in major bony fractures and defects. Stem cells can speed up bone repair in fractures/defects that would normally require extensive grafting, suggesting that mesenchymal stem cell use may provide a useful alternative to conventional grafting techniques. Treating tendon and ligament injuries in horses using stem cells, whether derived from adipose tissue or bone-marrow, has support in the veterinary literature. While further studies are necessary to fully characterize the use of cell-based therapeutics for treatment of bone fractures, stem cells are thought to mediate repair via five primary mechanisms: 1) providing an anti-inflammatory effect, 2) homing to damaged tissues and recruiting other cells, such as endothelial progenitor cells, that are necessary for tissue growth, 3) supporting tissue remodelling over scar formation, 4) inhibiting apoptosis, and 5) differentiating into bone, cartilage, tendon, and ligament tissue.

Significance of Stem cell Microenvironments

The microenvironment into which stem cells are transplanted significantly alters the capacity of engrafted cells for recovery and repair. The microenviroment provides growth factors and other chemical signals that guide appropriate differentiation of transplanted cell populations and direct transplanted cells to sites of trauma or disease.

Repair and recovery can then be mediated via three primary mechanisms: 1) formation and/or recruitment of new blood cells to the damaged region; 2) prevention of programed cell death or apoptosis; and 3) suppression of inflammation. To further enrich blood supply to the damaged areas, and consequently promote tissue regeneration, platelet-rich plasma could be used in conjunction with stem cell transplantation. The efficacy of some stem cell populations may also be affected by the method of delivery; for instance, to regenerate bone, stem cells are often introduced in a scaffold where they produce the minerals necessary for generation of functional bone.

Sources of Autologous (Patient-derived) Stem Cells

Autologous stem cells intended for regenerative therapy are generally isolated either from the patient's bone marrow or from adipose tissue. The number of stem cells transplanted into damaged tissue may alter efficacy of treatment. Accordingly, stem cells derived from bone marrow aspirates, for instance, are cultured in specialised laboratories for expansion to millions of cells. Although adipose-derived tissue also requires processing prior to use, the culturing methodology for adipose-derived stem cells is not as extensive as that for bone marrow-derived cells. While it is thought that bone-marrow derived stem cells are preferred for bone, cartilage, ligament, and tendon repair, others believe that the less challenging collection techniques and the multicellular microenvironment already present in adipose-derived stem cell fractions make the latter the preferred source for autologous transplantation.

Currently Available Treatments for Horses and Dogs Suffering From Orthopaedic Conditions

Autologous or allogeneic stem cells are currently used as an adjunctive therapy in the surgical repair of some types of fractures in dogs and horses. Autologous stem cell-based treatments for ligament injury, tendon injury, osteoarthritis, osteochondrosis, and sub-chondral bone cysts have been commercially available to practicing veterinarians to treat horses since 2003 in the United States and since 2006 in the United Kingdom. Autologous stem-cell based treatments for tendon injury, ligament injury, and osteoarthritis in dogs have been available to veterinarians in the United States since 2005. Over 3000 privately-owned horses and dogs have been treated with autologous adipose-derived stem cells. The efficacy of these treatments has been shown in double-blind clinical trials for dogs with osteoarthritis of the hip and elbow and horses with tendon damage.

The efficacy of using stem cells, whether adipose-derived or bone-marrow derived, for treating tendon and ligament injuries in horses has support in the veterinary literature.

Developments in Stem Cell Treatments in Veterinary Internal Medicine

Currently, research is being conducted to develop stem cell treatments for: 1) horses suffering from COPD, neurologic disease, and laminitis; and 2) dogs and cats suffering from heart disease, liver disease, kidney disease, neurologic disease, and immune-mediated disorders.

Embryonic Stem Cell Controversy

There is widespread controversy over the use of human embryonic stem cells. This controversy primarily targets the techniques used to derive new embryonic stem cell lines, which often requires the destruction of the blastocyst. Opposition to the use of human embryonic stem cells in research is often based on philosophical, moral or religious objections.

Stem Cell Treatments Around the World

China

Stem cell research and treatment is currently being practiced at a clinical level in the People's Republic of China. The Ministry of Health of the People's Republic of China has permitted the use of stem cell therapy for conditions beyond those approved of in Western countries such as the United States, United Kingdom, and Australia. The Western World has scrutinised China for its failed attempts to meet international documentation standards of these trials and procedures, despite the overwhelmingly positive anecdotal results.

Stem cell therapies provided in China utilize a variety of cell types including umbilical cord stem cells and olfactory ensheathing cells. The stem cells are then expanded in centralised blood banks before being used in stem cell treatments. State-funded companies based in the Shenzhen Hi-Tech Industrial Zone treat the symptoms of numerous disorders with adult stem cell therapy. Hospitals throughout eastern China provide numerous therapies to patients in coordination with the stem cell providers. These companies' therapies are currently focused on the treatment of neurodegenerative and cardiovascular disorders.

The most radical successes of Chinese adult stem cell therapy have been in treating the brain. These therapies administer stem cells directly to the brain to promote greater motor and brain function in patients with Cerebral Palsy, Alzheimer's, and brain injuries. However, retrospective studies have shown that Chinese use of fetal-derived brain tissue in spinal cord injured human subjects were not as promising as once thought: the phenotype and the fate of the transplanted cells, described as olfactory ensheathing cells, were unknown. As well, perioperative morbidity and lack of functional benefit were identified as the most serious clinical shortcomings. Furthermore, the extent of regulatory policy in the use of stem cell therapies in China is unclear. In the absence of a valid clinical trials protocol, and more regulatory oversight, Western regulatory agencies advise patients and physicians to be cautious when selecting Chinese stem cell therapeutic centres.

Mexico

Stem cell treatment is currently being practiced at a clinical level in Mexico. An International Health Department Permit (COFEPRIS) is required. This permit allows the use of stem cell types beyond those approved of in Western countries such as the United States or Europe. Stem cell therapies provided in Mexico utilize patient Adipose, Bone Marrow, or Donor Placenta sources.

South Korea

In 2005, South Korean scientists claimed to have generated stem cells that were tailored to match the recipient. Each of the 11 new stem cell lines was developed using somatic cell nuclear transfer (SCNT) technology. The resultant cells were thought to match the genetic material of the recipient, thus suggesting minimal to no cell rejection.

This study, however, was eventually discredited as the primary researcher, Dr. Woo Suk Hwang, admitted to using cells obtained from his research staff. In Dec 2005, claims were put forward that his research had been manipulated to wrongfully indicate positive results. Later that month, these claims were confirmed by an academic panel.

Ukraine

Today, Ukraine is permitted to perform clinical trials of stem cell treatments (Order of the MH of Ukraine [1] 630 "About carrying out clinical trials of stem cells", 2008) for the treatment of these pathologies: pancreatic necrosis, cirrhosis, hepatitis, burn disease, diabetes, multiple sclerosis, critical lower limb ischemia. The first medical institution granted the right to conduct clinical trials became the "Institute of Cell Therapy"(Kiev).

Cell Potency

The potency of a cell specifies its differentiation potential, or potential to differentiate into different cell types. Totipotency is the ability of a single cell to divide and produce all the differentiated cells in an organism, including extraembryonic tissues. Totipotent cells include spores and zygotes. In some organisms, cells can dedifferentiate and regain totipotency. For example, a plant cutting or callus can be used to grow an entire plant.

Human development begins when a sperm fertilizes an egg and creates a single totipotent cell (zygote). In the first hours after fertilization, this cell divides into identical totipotent cells, which can later develop into any of the three germ layers of a human (endoderm, mesoderm, or ectoderm) and into cells of the cytotrophoblast layer or syncytiotrophoblast layer of the placenta. After reaching the 16-cell stage, the totipotent cells of the morula differentiate into cells that will eventually become either the Blastocyst's Inner cell mass or outer trophoblasts. Approximately four days after fertilization and after several cycles of cell division, these totipotent cells begin to specialize. The inner cell mass, the source of embryonic stem cells, is pluripotent, not totipotent.

Research on *Caenorhabditis elegans* suggests that multiple mechanisms including RNA regulation may play a role in maintaining totipotency at different stages of development in some species.

Pluripotency

In cell biology, pluripotency (from the Latin "plurimus" or "very many" and "potentia" or "powered") refers to a stem cell that has the potential to differentiate into any of the three germ layers: endoderm (interior stomach lining, gastrointestinal tract, the lungs), mesoderm (muscle, bone, blood, urogenital), or ectoderm (epidermal tissues and nervous system). Pluripotent stem cells can give rise to any fetal or adult cell type. However, alone they cannot develop into a fetal or adult animal because they lack the potential to contribute to extraembryonic tissue, such as the placenta.

Progenitor Cell

A progenitor cell is a biological cell that, like a stem cell, has a tendency to differentiate into a specific type of cell, but is already more specific than a stem cell and is pushed to differentiate into its "target" cell. The most important difference between stem cells and progenitor cells is that stem cells can replicate indefinitely, whereas progenitor cells can only divide a limited number of times. Controversy about the exact definition remains and the concept is still evolving.

The terms "progenitor cell" and "stem cell" are sometimes equated.

Properties

Most progenitors are described as oligopotent. In this point of view, they may be compared to adult stem cells. But progenitors are said to be in a further stage of cell differentiation. They are in the "centre" between stem cells and fully differentiated cells. The kind of potency they have depends on the type of their "parent" stem cell and also on their niche. Like stem cells, mostly, they are formed and transported in a colony, with the right conditions for them to grow and differentiate into their target tissues. In contrast, some progenitor cells were found during research, and were isolated. After their marker was found, it was proven that these progenitor could move through the body and migrate towards the tissue where they are needed. Many properties are shared by adult stem cells and progenitor cells. But still, controversy remains because embryonic stem cells are true stem cells in that they are pluripotent and show unlimited capacity for self-renewal. In contrast, many cells termed adult stem cells would be better defined as progenitor cells, as their capacities for unlimited self-renewal and plasticity have not been comprehensively demonstrated.

Function

The majority of progenitor cells lie dormant or possess little activity in the tissue in which they reside. They exhibit slow growth and their main role is to replace cells lost by normal attrition. In case of tissue injury, damaged or dead cells, progenitor cells can be activated. Growth factors or cytokines are two substances which trigger the progenitors to mobilize towards the damaged tissue. At the same time, they start to differentiate into the target cells. Not all progenitors are mobile and are situated near the tissue of their target differentiation. When the cytokines, growth factors and other cell division enhancing stimulators take on the progenitors, a higher rate of cell division is introduced. It leads to the recovery of the tissue.

Examples

The characterisation or the defining principle of progenitor cells, in order to separate them from others, is based on the different cell markers rather than their morphological appearance.

- Satellite cells found in muscles. They play a major role in muscle cell differentiation and injury recoveries.
- Intermediate progenitor cells formed in the subventricular zone. Some of these transit amplifying neural progenitors migrate via rostral migratory stream to the olfactory bulb and differentiate further into specific types of neural cells.
- Bone marrow stromal cells, basal cell of epidermis have 10% of progenitor cell, although they are often classed as stem cells due to their high plasticity and potentially unlimited capacity for self renewal.
- Periosteum contains progenitor cells that develop into osteoblasts and chondroblasts.
- Pancreatic progenitor cells are among the most studied progenitors. They are used in research to develop a cure against diabetes type-1.
- Angioblasts or Endothelial progenitor cells (EPC). These are very important for research on fracture and wounds healing.

Development of the Human Cerebral Cortices

Before E40 (embryonic day 40), progenitor cells generate other progenitor cells; after that period progenitor cells produce only dissimilar daughters. The cells from a single progenitor cell form a proliferative unit that creates on cortical column; these columns contain a variety of neurons with different shapes.

Oligopotency

In biology, oligopotency is the ability of progenitor cells to differentiate into a few cell types. It is a degree of potency. Examples of oligopotent stem cells are the lymphoid or myeloid stem cells. Examples of progenitor cells are Vascular Stem Cells which have the capacity to become both Endothelial or smooth muscle cells.

Unipotency

In cell biology, a unipotent cell is one that has the capacity to develop/differentiate into only one type of tissue/cell type. The most common of these in humans are skin cells. This cell has a unique property: self-renewal. This property distinguishes it from most other terminally differentiated non-stem cells.

Hepatocytes, which constitute most of the Cytoplasmic mass of the human liver, are unipotent. The liver's ability to regenerate from as little as 25% of its original mass is attributed to this property. A close synonym for *unipotent cell* is *precursor cell.*

7

Focus on Chromosome, DNA, RNA, Nucleotide and Genome

Chromosome

A chromosome is an organized structure of DNA and protein that is found in cells. It is a single piece of coiled DNA containing many genes, regulatory elements and other nucleotide sequences. Chromosomes also contain DNA-bound proteins, which serve to package the DNA and control its functions. Chromosomes vary widely between different organisms. The DNA molecule may be circular or linear, and can be composed of 10,000 to 1,000,000,000 nucleotides in a long chain.

Typically, eukaryotic cells (cells with nuclei) have large linear chromosomes and prokaryotic cells (cells without defined nuclei) have smaller circular chromosomes, although there are many exceptions to this rule. Also, cells may contain more than one type of chromosome; for example, mitochondria in most eukaryotes and chloroplasts in plants have their own small chromosomes. In eukaryotes, nuclear chromosomes are packaged by proteins into a condensed structure called chromatin. This allows the very long DNA molecules to fit into the cell nucleus.

The structure of chromosomes and chromatin varies through the cell cycle. Chromosomes are the essential unit for cellular division and must be replicated, divided, and passed successfully to their daughter cells so as to ensure the genetic diversity and survival of their progeny. Chromosomes may exist as either duplicated or unduplicated. Unduplicated chromosomes are single linear strands, whereas duplicated chromosomes (copied during synthesis phase) contain two copies joined by a centromere. Compaction of the duplicated chromosomes during mitosis and meiosis results in the classic four-arm structure (pictured to the right). Chromosomal recombination plays a vital role in genetic diversity. If these structures are manipulated incorrectly, through processes known as chromosomal instability and translocation, the cell may undergo mitotic catastrophe and die, or it may unexpectedly evade apoptosis leading to the progression of cancer.

In practice "chromosome" is a rather loosely defined term. In prokaryotes and viruses, the term genophore is more appropriate when no chromatin is present. However, a large body of work uses the term chromosome regardless of chromatin content. In prokaryotes, DNA is usually arranged as a circle, which is tightly coiled in on itself, sometimes accompanied by one or more smaller, circular DNA molecules called plasmids. These small circular genomes are also found in mitochondria and chloroplasts, reflecting their bacterial origins. The simplest genophores are found in viruses: these DNA or RNA molecules are short linear or circular genophores that often lack structural proteins.

In a series of experiments beginning in the mid-1880s, Theodor Boveri gave the definitive demonstration that chromosomes are the vectors of heredity. His two principles were the *continuity* of chromosomes and the *individuality* of chromosomes. It is the second of these principles that was so original. Wilhelm Roux suggested that each chromosome carries a different genetic load. Boveri was able to test and confirm this hypothesis. Aided by the rediscovery at the start of the 1900s of Gregor Mendel's earlier work, Boveri was able to point out the connection between the rules of inheritance and the behaviour of the chromosomes. Boveri influenced two generations of American cytologists: Edmund Beecher Wilson, Walter Sutton and Theophilus Painter were all influenced by Boveri (Wilson and Painter actually worked with him). In his famous textbook *The Cell in Development and Heredity*, Wilson linked together the independent work of Boveri and Sutton (both around 1902) by naming the chromosome theory of inheritance the "Sutton-Boveri Theory" (the names are sometimes reversed). Ernst Mayr remarks that the theory was hotly contested by some famous geneticists: William Bateson, Wilhelm Johannsen, Richard Goldschmidt and T.H. Morgan, all of a rather dogmatic turn-of-mind. Eventually, complete proof came from chromosome maps in Morgan's own lab.

In Eukaryotes

Eukaryotes (cells with nuclei such as those found in plants, yeast, and animals) possess multiple large linear chromosomes contained in the cell's nucleus. Each chromosome has one centromere, with one or two arms projecting from the centromere, although, under most circumstances, these arms are not visible as such. In addition, most eukaryotes have a small circular mitochondrial genome, and some eukaryotes may have additional small circular or linear cytoplasmic chromosomes.

In the nuclear chromosomes of eukaryotes, the uncondensed DNA exists in a semi-ordered structure, where it is wrapped around histones (structural proteins), forming a composite material called chromatin.

Chromatin

Chromatin is the complex of DNA and protein found in the eukaryotic nucleus, which packages chromosomes. The structure of chromatin varies significantly between different stages of the cell cycle, according to the requirements of the DNA.

Interphase Chromatin

During interphase (the period of the cell cycle where the cell is not dividing), two types of chromatin can be distinguished:

- Euchromatin, which consists of DNA that is active, e.g., being expressed as protein.
- Heterochromatin, which consists of mostly inactive DNA. It seems to serve structural purposes during the chromosomal stages. Heterochromatin can be further distinguished into two types:
  - o *Constitutive heterochromatin*, which is never expressed. It is located around the centromere and usually contains repetitive sequences.
  - o *Facultative heterochromatin*, which is sometimes expressed.

Individual chromosomes cannot be distinguished at this stage – they appear in the nucleus as a homogeneous tangled mix of DNA and protein.

Metaphase Chromatin and Division

In the early stages of mitosis or meiosis (cell division), the chromatin strands become more and more condensed. They cease to function as accessible genetic material (transcription stops) and become a compact transportable form. This compact form makes the individual chromosomes visible, and they form the classic four arm structure, a pair of sister chromatids attached to each other at the centromere. The shorter arms are called *p arms* (from the French *petit*, small) and the longer arms are called *q arms* (*q* follows *p* in the Latin alphabet). This is the only natural context in which individual chromosomes are visible with an optical microscope.

During divisions, long microtubules attach to the centromere and the two opposite ends of the cell. The microtubules then pull the chromatids apart, so that each daughter cell inherits one set of chromatids. Once the cells have divided, the chromatids are uncoiled and can function again as chromatin. In spite of their appearance, chromosomes are structurally highly condensed, which enables these giant DNA structures to be contained within a cell nucleus. The self-assembled microtubules form the spindle, which attaches to chromosomes at specialized structures called kinetochores, one of which is present on each sister chromatid. A special DNA base sequence in the region of the kinetochores provides, along with special proteins, longer-lasting attachment in this region.

In Humans

Chromosomes can be divided into two types—autosomes, and sex chromosomes. Certain genetic traits are linked to your sex, and are passed on through the sex chromosomes. The autosomes contain the rest of the genetic hereditary information. All act in the same way during cell division. Human cells have 23 pairs of large linear nuclear chromosomes, (22 pairs of autosomes and one pair of sex chromosomes) giving a total of 46 per cell. In addition to these, human cells have many hundreds of copies of the mitochondrial genome. Sequencing of the human genome has provided a great deal of information about each of the chromosomes. Below is a table compiling statistics for the chromosomes, based on the Sanger Institute's human genome information in the Vertebrate Genome Annotation (VEGA) database. Number of genes is an estimate as it is in part based on gene predictions. Total chromosome length is an estimate as well, based on the estimated size of unsequenced heterochromatin regions.

Chromosome	*Genes*	*Total bases*	*Sequenced bases*
1	4,220	247,199,719	224,999,719
2	1,491	242,751,149	237,712,649
3	1,550	199,446,827	194,704,827
4	446	191,263,063	187,297,063
5	609	180,837,866	177,702,766
6	2,281	170,896,993	167,273,993
7	2,135	158,821,424	154,952,424
8	1,106	146,274,826	142,612,826
9	1,920	140,442,298	120,312,298
10	1,793	135,374,737	131,624,737
11	379	134,452,384	131,130,853
12	1,430	132,289,534	130,303,534
13	924	114,127,980	95,559,980
14	1,347	106,360,585	88,290,585
15	921	100,338,915	81,341,915
16	909	88,822,254	78,884,754
17	1,672	78,654,742	77,800,220

18	519	76,117,153	74,656,155
19	1,555	63,806,651	55,785,651
20	1,008	62,435,965	59,505,254
21	578	46,944,323	34,171,998
22	1,092	49,528,953	34,893,953
X (sex chromosome)	1,846	154,913,754	151,058,754
Y (sex chromosome)	454	57,741,652	25,121,652
Total	32,185	3,079,843,747	2,857,698,560

In prokaryotes

The prokaryotes – bacteria and archaea – typically have a single circular chromosome, but many variations do exist. Most bacteria have a single circular chromosome that can range in size from only 160,000 base pairs in the endosymbiotic bacterium *Candidatus Carsonella ruddii*, to 12,200,000 base pairs in the soil-dwelling bacterium *Sorangium cellulosum*. Spirochaetes of the genus *Borrelia* are a notable exception to this arrangement, with bacteria such as *Borrelia burgdorferi*, the cause of Lyme disease, containing a single linear chromosome.

Structure in Sequences

Prokaryotic chromosomes have less sequence-based structure than eukaryotes. Bacteria typically have a single point (the origin of replication) from which replication starts, whereas some archaea contain multiple replication origins. The genes in prokaryotes are often organized in operons, and do not usually contain introns, unlike eukaryotes.

DNA Packaging

Prokaryotes do not possess nuclei. Instead, their DNA is organized into a structure called the nucleoid. The nucleoid is a distinct structure and occupies a defined region of the bacterial cell. This structure is, however, dynamic and is maintained and remodeled by the actions of a range of histone-like proteins, which associate with the bacterial chromosome. In archaea, the DNA in chromosomes is even more organized, with the DNA packaged within structures similar to eukaryotic nucleosomes. Bacterial chromosomes tend to be tethered to the plasma membrane of the bacteria. In molecular biology application, this allows for its isolation from plasmid DNA by centrifugation of lysed bacteria and pelleting of the membranes (and the attached DNA).

Number of Chromosomes in Various Organisms

Normal members of a particular eukaryotic species all have the same number of nuclear chromosomes. Other eukaryotic chromosomes, i.e., mitochondrial and plasmid-like small chromosomes, are much more variable in number, and there may be thousands of copies per cell. Asexually reproducing species have one set of chromosomes, which are the same in all body cells. However, asexual species can be either haploid or diploid.

Sexually reproducing species have somatic cells (body cells), which are diploid [2n] having two sets of chromosomes, one from the mother and one from the father. Gametes, reproductive cells, are haploid [n]: They have one set of chromosomes. Gametes are produced by meiosis of a diploid germ line cell. During meiosis, the matching chromosomes of father and mother can exchange small parts of themselves (crossover), and thus create new chromosomes that are not inherited solely from either parent. When a male and a female gamete merge (fertilization), a new diploid organism is formed.

Some animal and plant species are polyploid [Xn]: They have more than two sets of homologous chromosomes. Plants important in agriculture such as tobacco or wheat are often polyploid, compared to their ancestral species. Wheat has a haploid number of seven chromosomes, still seen in some cultivars as well as the wild progenitors. The more-common pasta and bread wheats are polyploid, having 28 (tetraploid) and 42 (hexaploid) chromosomes, compared to the 14 (diploid) chromosomes in the wild wheat.

Prokaryotes

Prokaryote species generally have one copy of each major chromosome, but most cells can easily survive with multiple copies. For example, *Buchnera*, a symbiont of aphids has multiple copies of its chromosome, ranging from 10–400 copies per cell. However, in some large bacteria, such as *Epulopiscium fishelsoni* up to 100,000 copies of the chromosome can be present. Plasmids and plasmid-like small chromosomes are, as in eukaryotes, very variable in copy number. The number of plasmids in the cell is almost entirely determined by the rate of division of the plasmid – fast division causes high copy number, and vice versa.

Karyotype

In general, the karyotype is the characteristic chromosome complement of a eukaryote species. The preparation and study of karyotypes is part of cytogenetics. Although the replication and transcription of DNA is highly standardized in eukaryotes, *the same cannot be said for their karyotypes*, which are often highly variable. There may be variation between species in chromosome number and in detailed organization. In some cases, there is significant variation within species. Often there is:

1. variation between the two sexes
2. variation between the germ-line and soma (between gametes and the rest of the body)

3. variation between members of a population, due to balanced genetic polymorphism
4. geographical variation between races
5. mosaics or otherwise abnormal individuals.

Historical Note

Investigation into the human karyotype took many years to settle the most basic question. How many chromosomes does a normal diploid human cell contain? In 1912, Hans von Winiwarter reported 47 chromosomes in spermatogonia and 48 in oogonia, concluding an XX/XO sex determination mechanism. Painter in 1922 was not certain whether the diploid number of man is 46 or 48, at first favouring 46. He revised his opinion later from 46 to 48, and he correctly insisted on humans having an XX/XY system.

New techniques were needed to definitively solve the problem:

1. Using cells in culture
2. Pretreating cells in a hypotonic solution, which swells them and spreads the chromosomes
3. Arresting mitosis in metaphase by a solution of colchicine
4. Squashing the preparation on the slide forcing the chromosomes into a single plane
5. Cutting up a photomicrograph and arranging the result into an indisputable karyogram.

It took until the mid-1950s for it to become generally accepted that the human karyotype include only 46 chromosomes. Considering the techniques of Winiwarter and Painter, their results were quite remarkable. Chimpanzees (the closest living relatives to modern humans) have 48 chromosomes.

Aberrations

Chromosomal aberrations are disruptions in the normal chromosomal content of a cell and are a major cause of genetic conditions in humans, such as Down syndrome. Some chromosome abnormalities do not cause disease in carriers, such as translocations, or chromosomal inversions, although they may lead to a higher chance of birthing a child with a chromosome disorder. Abnormal numbers of chromosomes or chromosome sets, aneuploidy, may be lethal or give rise to genetic disorders. Genetic counseling is offered for families that may carry a chromosome rearrangement. The gain or loss of DNA from chromosomes can lead to a variety of genetic disorders. Human examples include:

- Cri du chat, which is caused by the deletion of part of the short arm of chromosome 5. "Cri du chat" means "cry of the cat" in French, and the condition was so-named because affected babies make high-pitched cries that sound like those of a cat. Affected individuals have wide-set eyes, a small head and jaw, moderate to severe mental health issues, and are very short.

- Down syndrome, usually is caused by an extra copy of chromosome 21 (trisomy 21). Characteristics include decreased muscle tone, stockier build, asymmetrical skull, slanting eyes and mild to moderate developmental disability.
- Edwards syndrome, which is the second-most-common trisomy; Down syndrome is the most common. It is a trisomy of chromosome 18. Symptoms include motor retardation, developmental disability and numerous congenital anomalies causing serious health problems. Ninety percent die in infancy; however, those that live past their first birthday usually are quite healthy thereafter. They have a characteristic clenched hands and overlapping fingers.
- Idic15, abbreviation for Isodicentric 15 on chromosome 15; also called the following names due to various researches, but they all mean the same; IDIC(15), Inverted duplication 15, extra Marker, Inv dup 15, partial tetrasomy 15
- Jacobsen syndrome, also called the terminal 11q deletion disorder. This is a very rare disorder. Those affected have normal intelligence or mild developmental disability, with poor expressive language skills. Most have a bleeding disorder called Paris-Trousseau syndrome.
- Klinefelter's syndrome (XXY). Men with Klinefelter syndrome are usually sterile, and tend to have longer arms and legs and to be taller than their peers. Boys with the syndrome are often shy and quiet, and have a higher incidence of speech delay and dyslexia. During puberty, without testosterone treatment, some of them may develop gynecomastia.
- Patau Syndrome, also called D-Syndrome or trisomy-13. Symptoms are somewhat similar to those of trisomy-18, but they do not have the characteristic hand shape.
- Small supernumerary marker chromosome. This means there is an extra, abnormal chromosome. Features depend on the origin of the extra genetic material. Cat-eye syndrome and isodicentric chromosome 15 syndrome (or Idic15) are both caused by a supernumerary marker chromosome, as is Pallister-Killian syndrome.
- Triple-X syndrome (XXX). XXX girls tend to be tall and thin. They have a higher incidence of dyslexia.
- Turner syndrome (X instead of XX or XY). In Turner syndrome, female sexual characteristics are present but underdeveloped. People with

Turner syndrome often have a short stature, low hairline, abnormal eye features and bone development and a "caved-in" appearance to the chest.

- XYY syndrome. XYY boys are usually taller than their siblings. Like XXY boys and XXX girls, they are somewhat more likely to have learning difficulties.
- Wolf-Hirschhorn syndrome, which is caused by partial deletion of the short arm of chromosome 4. It is characterized by severe growth retardation and severe to profound mental health issues.

Chromosomal mutations produce changes in whole chromosomes (more than one gene) or in the number of chromosomes present.

- Deletion – loss of part of a chromosome
- Duplication – extra copies of a part of a chromosome
- Inversion – reverse the direction of a part of a chromosome
- Translocation – part of a chromosome breaks off and attaches to another chromosome

Most mutations are neutral – have little or no effect. Chromosomal aberrations are the changes in the structure of chromosomes. It has a great role in evolution. A detailed graphical display of all human chromosomes and the diseases annotated at the correct spot may be found at the Oak Ridge National Laboratory.

DNA

Deoxyribonucleic acid or DNA, is a nucleic acid that contains the genetic instructions used in the development and functioning of all known living organisms (with the exception of RNA viruses). The main role of DNA molecules is the long-term storage of information. DNA is often compared to a set of blueprints, like a recipe or a code, since it contains the instructions needed to construct other components of cells, such as proteins and RNA molecules. The DNA segments that carry this genetic information are called genes, but other DNA sequences have structural purposes, or are involved in regulating the use of this genetic information.

DNA consists of two long polymers of simple units called nucleotides, with backbones made of sugars and phosphate groups joined by ester bonds. These two strands run in opposite directions to each other and are therefore anti-parallel. Attached to each sugar is one of four types of molecules called bases. It is the sequence of these four bases along the backbone that encodes information. This information is read using the genetic code, which specifies the sequence of the amino acids within proteins. The code is read by copying stretches of DNA into the related nucleic acid RNA, in a process called transcription. Within cells, DNA is organized into long structures called chromosomes. These chromosomes are duplicated before cells divide, in a process called DNA replication. Eukaryotic organisms (animals, plants, fungi, and protists) store most of their DNA inside the cell nucleus and some of their DNA in organelles, such as mitochondria or chloroplasts. In contrast, prokaryotes (bacteria and archaea) store their DNA only in the cytoplasm. Within the chromosomes, chromatin proteins such as histones compact and organize DNA. These compact structures guide the interactions between DNA and other proteins, helping control which parts of the DNA are transcribed.

Properties

DNA is a long polymer made from repeating units called nucleotides. As first discovered by James D. Watson and Francis Crick, the structure of DNA of all species comprises two helical chains each coiled round the same axis, and each with a pitch of 34 Ångströms (3.4 nanometres) and a radius of 10 Ångströms (1.0 nanometres). According to another study, when measured in a particular solution, the DNA chain measured 22 to 26 Ångströms wide (2.2 to 2.6 nanometres), and one nucleotide unit measured 3.3 Å (0.33 nm) long. Although each individual repeating unit is very small, DNA polymers can be very large molecules containing millions of nucleotides. For instance, the largest human chromosome, chromosome number 1, is approximately 220 million base pairs long.

In living organisms, DNA does not usually exist as a single molecule, but instead as a pair of molecules that are held tightly together. These two long strands entwine like vines, in the shape of a double helix. The nucleotide repeats contain both the segment of the backbone of the molecule, which holds the chain together, and a base, which interacts with the other DNA strand in the helix. A base linked to a sugar is called a nucleoside and a base linked to a sugar and one or more phosphate groups is called a nucleotide. If multiple nucleotides are linked together, as in DNA, this polymer is called a polynucleotide.

The backbone of the DNA strand is made from alternating phosphate and sugar residues. The sugar in DNA is 2-deoxyribose, which is a pentose (five-carbon) sugar. The sugars are joined together by phosphate groups that form phosphodiester bonds between the third and fifth carbon atoms of adjacent sugar rings. These asymmetric bonds mean a strand of DNA has a direction. In a double helix the direction of the nucleotides in one strand is opposite to their direction in the other strand: the strands are *antiparallel*. The asymmetric ends of DNA strands are called the 52 (*five prime*) and 32 (*three prime*) ends, with the 5' end having a terminal phosphate group and the 3' end a terminal hydroxyl group. One major difference between DNA and RNA is the sugar, with the 2-deoxyribose in DNA being replaced by the alternative pentose sugar ribose in RNA.

The DNA double helix is stabilized primarily by two forces: hydrogen bonds between nucleotides and base-stacking interactions among the aromatic bases. In the aqueous environment of the cell, the conjugated ð bonds of nucleotide bases align perpendicular to the axis of the DNA molecule, minimizing their interaction with the solvation shell and therefore, the Gibbs free energy. The four bases found in DNA are adenine (abbreviated A), cytosine (C), guanine (G) and thymine (T). These four bases are attached to the sugar/ phosphate to form the complete nucleotide, as shown for adenosine monophosphate. These bases are classified into two types; adenine and guanine are fused five- and six-membered heterocyclic compounds called purines, while cytosine and thymine are six-membered rings called pyrimidines. A fifth pyrimidine base, called uracil (U), usually takes the place of thymine in RNA and differs from thymine by lacking a methyl group on its ring. Uracil is not usually found in DNA, occurring only as a breakdown product of cytosine. In addition to RNA and DNA, a large number of artificial nucleic acid analogues have also been created to study the proprieties of nucleic acids, or for use in biotechnology.

Grooves

Twin helical strands form the DNA backbone. Another double helix may be found by tracing the spaces, or grooves, between the strands. These voids are adjacent to the base pairs and may provide a binding site. As the strands are not directly opposite each other, the grooves are unequally sized. One groove, the major groove, is 22 Å wide and the other, the minor groove, is 12 Å wide. The narrowness of the minor groove means that the edges of the bases are more accessible in the major groove. As a result, proteins like transcription factors that can bind to specific sequences in double-stranded DNA usually make contacts to the sides of the bases exposed in the major groove. This situation varies in unusual conformations of DNA within the cell, but the major and minor grooves are always named to reflect the differences in size that would be seen if the DNA is twisted back into the ordinary B form.

Base Pairing

Each type of base on one strand forms a bond with just one type of base on the other strand. This is called complementary base pairing. Here, purines form hydrogen bonds to pyrimidines, with A bonding only to T, and C bonding only to G. This arrangement of two nucleotides binding together across the double helix is called a base pair. As hydrogen bonds are not covalent, they can be broken and rejoined relatively easily. The two strands of DNA in a double helix can therefore be pulled apart like a zipper, either by a mechanical force or high temperature. As a result of this complementarity, all the information in the double-stranded sequence of a DNA helix is duplicated on each strand, which is vital in DNA replication. Indeed, this reversible and specific interaction between complementary base pairs is critical for all the functions of DNA in living organisms.

The two types of base pairs form different numbers of hydrogen bonds, AT forming two hydrogen bonds, and GC forming three hydrogen bonds. DNA with high GC-content is more stable than DNA with low GC-content, due to the added stability of an additional hydrogen bond. As a result, it is both the percentage of GC base pairs and the overall length of a DNA double helix that determine the strength of the association between the two strands of DNA. Long DNA helices with a high GC content have stronger-interacting strands, while short helices with high AT content have weaker-interacting strands. In biology, parts of the DNA double helix that need to separate easily, such as the TATAAT Pribnow box in some promoters, tend to have a high AT content, making the strands easier to pull apart.

In the laboratory, the strength of this interaction can be measured by finding the temperature required to break the hydrogen bonds, their melting temperature (also called T_m value). When all the base pairs in a DNA double helix melt, the strands separate and exist in solution as two entirely independent molecules. These single-stranded DNA molecules (*ssDNA*) have no single common shape, but some conformations are more stable than others.

Sense and Antisense

A DNA sequence is called "sense" if its sequence is the same as that of a messenger RNA copy that is translated into protein. The sequence on the opposite strand is called the "antisense" sequence. Both sense and antisense sequences can exist on different parts of the same strand of DNA (i.e. both strands contain both sense and antisense sequences). In both prokaryotes and eukaryotes, antisense RNA sequences are produced, but the functions of these RNAs are not entirely clear. One proposal is that antisense RNAs are involved in regulating gene expression through RNA-RNA base pairing.

A few DNA sequences in prokaryotes and eukaryotes, and more in plasmids and viruses, blur the distinction between sense and antisense strands by having overlapping genes. In these cases, some DNA sequences do double duty, encoding one protein when read along one strand, and a second protein when read in the opposite direction along the other strand. In bacteria, this overlap may be involved in the regulation of gene transcription, while in viruses, overlapping genes increase the amount of information that can be encoded within the small viral genome.

Supercoiling

DNA can be twisted like a rope in a process called DNA supercoiling. With DNA in its "relaxed" state, a strand usually circles the axis of the double helix once every 10.4 base pairs, but if the DNA is twisted the strands become more tightly or more loosely wound. If the DNA is twisted in the direction of the helix, this is positive supercoiling, and the bases are held more tightly together. If they are twisted in the opposite direction, this is negative supercoiling, and the bases come apart more easily. In nature, most DNA has slight negative supercoiling that is introduced by enzymes called topoisomerases. These enzymes are also needed to relieve the twisting stresses introduced into DNA strands during processes such as transcription and DNA replication.

Alternate DNA Structures

DNA exists in many possible conformations that include A-DNA, B-DNA, and Z-DNA forms, although, only B-DNA and Z-DNA have been directly observed in functional organisms. The conformation that DNA adopts depends on the hydration level, DNA sequence, the amount and direction of supercoiling, chemical modifications of the bases, the type and concentration of metal ions, as well as the presence of polyamines in solution. The first published reports of A-DNA X-ray diffraction patterns— and also B-DNA used analyses based on Patterson transforms that provided only a limited amount of structural information for oriented fibers of DNA. An alternate analysis was then proposed by Wilkins *et al.*, in 1953, for the *in vivo* B-DNA X-ray diffraction/scattering patterns of highly hydrated DNA fibers in terms of squares of Bessel functions. In the same journal, James D. Watson and Francis Crick presented their molecular modeling analysis of the DNA X-ray diffraction patterns to suggest that the structure was a double-helix.

Although the 'B-DNA form' is most common under the conditions found in cells, it is not a well-defined conformation but a family of related DNA conformations that occur at the high hydration levels present in living cells. Their corresponding X-ray diffraction and scattering patterns are characteristic of molecular paracrystals with a significant degree of disorder. Compared to B-DNA, the A-DNA form is a wider right-handed spiral, with a shallow, wide minor groove and a narrower, deeper major groove. The A form occurs under non-physiological conditions in partially dehydrated samples of DNA, while in the cell it may be produced in hybrid pairings of DNA and RNA strands, as well as in enzyme-DNA complexes. Segments of DNA where the bases have been chemically modified by methylation may undergo a larger change in conformation and adopt the Z form. Here, the strands turn about the helical axis in a left-handed spiral, the opposite of the more common B form. These unusual structures can be recognized by specific Z-DNA binding proteins and may be involved in the regulation of transcription.

Alternate DNA Chemistry

For a number of years exobiologists have proposed the existence of a shadow biosphere, a postulated microbial biosphere of Earth that uses radically different biochemical and molecular processes than currently known life. One of the proposals was the existence of lifeforms that use arsenic instead of phosphorus in DNA.

A December 2010 NASA press conference revealed that the bacterium GFAJ-1, which has evolved in an arsenic-rich environment, is the first terrestrial lifeform found which may have this ability. The bacterium was found in Mono Lake, east of Yosemite National Park. GFAJ-1 is a rod-shaped extremophile bacterium in the family Halomonadaceae that, when starved of phosphorus, may be capable of incorporating the usually poisonous element arsenic in its DNA.

This discovery may lend weight to the long-standing idea that extraterrestrial life could have a different chemical makeup from life on Earth. The research was carried out by a team led by Felisa Wolfe-Simon, a geomicrobiologist and geobiochemist, a Postdoctoral Fellow of the NASA Astrobiology Institute with Arizona State University. This finding has, however, faced strong criticism from the scientific community; scientists have argued that there is no evidence that arsenic is actually incorporated into biomolecules. Independent conformation of this finding has also not yet been possible.

Quadruplex Structures

At the ends of the linear chromosomes are specialized regions of DNA called telomeres. The main function of these regions is to allow the cell to replicate chromosome ends using the enzyme telomerase, as the enzymes that normally replicate DNA cannot copy the extreme 32 ends of chromosomes. These specialized chromosome caps also help protect the DNA ends, and stop the DNA repair systems in the cell from treating them as damage to be corrected. In human cells, telomeres are usually lengths of single-stranded DNA containing several thousand repeats of a simple TTAGGG sequence.

These guanine-rich sequences may stabilize chromosome ends by forming structures of stacked sets of four-base units, rather than the usual base pairs found in other DNA molecules. Here, four guanine bases form a flat plate and these flat four-base units then stack on top of each other, to form a stable *G-quadruplex* structure. These structures are stabilized by hydrogen bonding between the edges of the bases and chelation of a metal ion in the centre of each four-base unit. Other structures can also be formed, with the central set of four bases coming from either a single strand folded around the bases, or several different parallel strands, each contributing one base to the central structure.

In addition to these stacked structures, telomeres also form large loop structures called telomere loops, or T-loops. Here, the single-stranded DNA curls around in a long circle stabilized by telomere-binding proteins. At the very end of the T-loop, the single-stranded telomere DNA is held onto a region of double-stranded DNA by the telomere strand disrupting the double-helical DNA and base pairing to one of the two strands. This triple-stranded structure is called a displacement loop or D-loop.

Branched DNA

In DNA fraying occurs when non-complementary regions exist at the end of an otherwise complementary double-strand of DNA. However, branched DNA can occur if a third strand of DNA is introduced and contains adjoining regions able to hybridize with the frayed regions of the pre-existing double-strand. Although the simplest example of branched DNA involves only three strands of DNA, complexes involving additional strands and multiple branches are also possible. Branched DNA can be used in nanotechnology to construct geometric shapes, see the section on uses in technology below.

Vibration

DNA may carry out low-frequency collective motion as observed by the Raman spectroscopy and analyzed with a quasi-continuum model.

Chemical Modifications

Base Modifications

The expression of genes is influenced by how the DNA is packaged in chromosomes, in a structure called chromatin. Base modifications can be involved in packaging, with regions that have low or no gene expression usually containing high levels of methylation of cytosine bases. For example, cytosine methylation, produces 5-methylcytosine, which is important for X-chromosome inactivation. The average level of methylation varies between organisms - the worm *Caenorhabditis elegans* lacks cytosine methylation, while vertebrates have higher levels, with up to 1% of their DNA containing 5-methylcytosine. Despite the importance of 5-methylcytosine, it can deaminate to leave a thymine base, so methylated cytosines are particularly prone to mutations. Other base modifications include adenine methylation in bacteria, the presence of 5-hydroxymethylcytosine in the brain, and the glycosylation of uracil to produce the "J-base" in kinetoplastids.

Damage

DNA can be damaged by many sorts of mutagens, which change the DNA sequence. Mutagens include oxidizing agents, alkylating agents and also high-energy electromagnetic radiation such as ultraviolet light and X-rays. The type of DNA damage produced depends on the type of mutagen. For example, UV light can damage DNA by producing thymine dimers, which are cross-links between pyrimidine bases. On the other hand, oxidants such as free radicals or hydrogen peroxide produce multiple forms of damage, including base modifications, particularly of guanosine, and double-strand breaks. A typical human cell contains about 150,000 bases that have suffered oxidative damage. Of these oxidative lesions, the most dangerous are double-strand breaks, as these are difficult to repair and can produce point mutations, insertions and deletions from the DNA sequence, as well as chromosomal translocations.

Many mutagens fit into the space between two adjacent base pairs, this is called *intercalation*. Most intercalators are aromatic and planar molecules; examples include ethidium bromide, daunomycin, and doxorubicin. In order for an intercalator to fit between base pairs, the bases must separate, distorting the DNA strands by unwinding of the double helix. This inhibits both transcription and DNA replication, causing toxicity and mutations. As a result, DNA intercalators are often carcinogens, and benzo[*a*]pyrene diol epoxide, acridines, aflatoxin and ethidium bromide are well-known examples. Nevertheless, due to their ability to inhibit DNA transcription and replication, other similar toxins are also used in chemotherapy to inhibit rapidly growing cancer cells.

Biological Functions

DNA usually occurs as linear chromosomes in eukaryotes, and circular chromosomes in prokaryotes. The set of chromosomes in a cell makes up its genome; the human genome has approximately 3 billion base pairs of DNA arranged into 46 chromosomes. The information carried by DNA is held in the sequence of pieces of DNA called genes. Transmission of genetic information in genes is achieved via complementary base pairing. For example, in transcription, when a cell uses the information in a gene, the DNA sequence is copied into a complementary RNA sequence through the attraction between the DNA and the correct RNA nucleotides. Usually, this RNA copy is then used to make a matching protein sequence in a process called translation, which depends on the same interaction between RNA nucleotides. In alternative fashion, a cell may simply copy its genetic information in a process called DNA replication. The details of these functions are covered in other articles; here we focus on the interactions between DNA and other molecules that mediate the function of the genome.

Genes and Genomes

Genomic DNA is tightly and orderly packed in the process called DNA condensation to fit the small available volumes of the cell. In eukaryotes, DNA is located in the cell nucleus, as well as small amounts in mitochondria and chloroplasts. In prokaryotes, the DNA is held within an irregularly shaped body in the cytoplasm called the nucleoid. The genetic information in a genome is held within genes, and the complete set of this information in an organism is called its genotype. A gene is a unit of heredity and is a region of DNA that influences a particular characteristic in an organism. Genes contain an open reading frame that can be transcribed, as well as regulatory sequences such as promoters and enhancers, which control the transcription of the open reading frame.

In many species, only a small fraction of the total sequence of the genome encodes protein. For example, only about 1.5% of the human genome consists of protein-coding exons, with over 50% of human DNA consisting of non-coding repetitive sequences. The reasons for the presence of so much noncoding DNA in eukaryotic genomes and the extraordinary differences in genome size, or *C-value*, among species represent a long-standing puzzle known as the "C-value enigma". However, DNA sequences that do not code protein may still encode functional non-coding RNA molecules, which are involved in the regulation of gene expression.

Some noncoding DNA sequences play structural roles in chromosomes. Telomeres and centromeres typically contain few genes, but are important for the function and stability of chromosomes. An abundant form of noncoding DNA in humans are pseudogenes, which are copies of genes that have been disabled by mutation. These sequences are usually just molecular fossils, although they can occasionally serve as raw genetic material for the creation of new genes through the process of gene duplication and divergence.

Transcription and Translation

A gene is a sequence of DNA that contains genetic information and can influence the phenotype of an organism. Within a gene, the sequence of bases along a DNA strand defines a messenger RNA sequence, which then defines one or more protein sequences. The relationship between the nucleotide sequences of genes and the amino-acid sequences of proteins is determined by the rules of translation, known collectively as the genetic code. The genetic code consists of three-letter 'words' called *codons* formed from a sequence of three nucleotides (e.g. ACT, CAG, TTT).

In transcription, the codons of a gene are copied into messenger RNA by RNA polymerase. This RNA copy is then decoded by a ribosome that reads the RNA sequence by base-pairing the messenger RNA to transfer RNA, which carries amino acids. Since there are 4 bases in 3-letter combinations, there are 64 possible codons (4^3 combinations). These encode the twenty standard amino acids, giving most amino acids more than one possible codon. There are also three 'stop' or 'nonsense' codons signifying the end of the coding region; these are the TAA, TGA and TAG codons.

Replication

Cell division is essential for an organism to grow, but, when a cell divides, it must replicate the DNA in its genome so that the two daughter cells have the same genetic information as their parent. The double-stranded structure of DNA provides a simple mechanism for DNA replication. Here, the two strands are separated and then each strand's complementary DNA sequence is recreated by an enzyme called DNA polymerase. This enzyme makes the complementary strand by finding the correct base through complementary base pairing, and bonding it onto the original strand. As DNA polymerases can only extend a DNA strand in a 52 to 32 direction, different mechanisms are used to copy the antiparallel strands of the double helix. In this way, the base on the old strand dictates which base appears on the new strand, and the cell ends up with a perfect copy of its DNA.

Interactions with Proteins

All the functions of DNA depend on interactions with proteins. These protein interactions can be non-specific, or the protein can bind specifically to a single DNA sequence. Enzymes can also bind to DNA and of these, the polymerases that copy the DNA base sequence in transcription and DNA replication are particularly important.

DNA-Binding Proteins

Structural proteins that bind DNA are well-understood examples of non-specific DNA-protein interactions. Within chromosomes, DNA is held in complexes with structural proteins. These proteins organize the DNA into a compact structure called chromatin. In eukaryotes this structure involves DNA binding to a complex of small basic proteins called histones, while in prokaryotes multiple types of proteins are involved. The histones form a disk-shaped complex called a nucleosome, which contains two complete turns of double-stranded DNA wrapped around its surface.

These non-specific interactions are formed through basic residues in the histones making ionic bonds to the acidic sugar-phosphate backbone of the DNA, and are therefore largely independent of the base sequence. Chemical modifications of these basic amino acid residues include methylation, phosphorylation and acetylation. These chemical changes alter the strength of the interaction between the DNA and the histones, making the DNA more or less accessible to transcription factors and changing the rate of transcription. Other non-specific DNA-binding proteins in chromatin include the high-mobility group proteins, which bind to bent or distorted DNA. These proteins are important in bending arrays of nucleosomes and arranging them into the larger structures that make up chromosomes. A distinct group of DNA-binding proteins are the DNA-binding proteins that specifically bind single-stranded DNA. In humans, replication protein A is the best-understood member of this family and is used in processes where the double helix is separated, including DNA replication, recombination and DNA repair. These binding proteins seem to stabilize single-stranded DNA and protect it from forming stem-loops or being degraded by nucleases.

In contrast, other proteins have evolved to bind to particular DNA sequences. The most intensively studied of these are the various transcription factors, which are proteins that regulate transcription. Each transcription factor binds to one particular set of DNA sequences and activates or inhibits the transcription of genes that have these sequences close to their promoters. The transcription factors do this in two ways. Firstly, they can bind the RNA polymerase responsible for transcription, either directly or through other mediator proteins; this locates the polymerase at the promoter and allows it to begin transcription. Alternatively, transcription factors can bind enzymes that modify the histones at the promoter; this will change the accessibility of the DNA template to the polymerase.

As these DNA targets can occur throughout an organism's genome, changes in the activity of one type of transcription factor can affect thousands of genes. Consequently, these proteins are often the targets of the signal transduction processes that control responses to environmental changes or cellular differentiation and development. The specificity of these transcription factors' interactions with DNA come from the proteins making multiple contacts to the edges of the DNA bases, allowing them to "read" the DNA sequence. Most of these base-interactions are made in the major groove, where the bases are most accessible.

DNA-Modifying Enzymes

Nucleases and Ligases

Nucleases are enzymes that cut DNA strands by catalyzing the hydrolysis of the phosphodiester bonds. Nucleases that hydrolyse nucleotides from the ends of DNA strands are called exonucleases, while endonucleases cut within strands. The most frequently used nucleases in molecular biology are the restriction endonucleases, which cut DNA at specific sequences. For instance, the EcoRV enzyme shown to the left recognizes the 6-base sequence 52 - GAT|ATC-32 and makes a cut at the vertical line. In nature, these enzymes protect bacteria against phage infection by digesting the phage DNA when it enters the bacterial cell, acting as part of the restriction modification system. In technology, these sequence-specific nucleases are used in molecular cloning and DNA fingerprinting.

Enzymes called DNA ligases can rejoin cut or broken DNA strands. Ligases are particularly important in lagging strand DNA replication, as they join together the short segments of DNA produced at the replication fork into a complete copy of the DNA template. They are also used in DNA repair and genetic recombination.

Topoisomerases and Helicases

Topoisomerases are enzymes with both nuclease and ligase activity. These proteins change the amount of supercoiling in DNA. Some of these enzymes work by cutting the DNA helix and allowing one section to rotate, thereby reducing its level of supercoiling; the enzyme then seals the DNA break. Other types of these enzymes are capable of cutting one DNA helix and then passing a second strand of DNA through this break, before rejoining the helix. Topoisomerases are required for many processes involving DNA, such as DNA replication and transcription. Helicases are proteins that are a type of molecular motor. They use the chemical energy in nucleoside triphosphates, predominantly ATP, to break hydrogen bonds between bases and unwind the DNA double helix into single strands. These enzymes are essential for most processes where enzymes need to access the DNA bases.

Polymerases

Polymerases are enzymes that synthesize polynucleotide chains from nucleoside triphosphates. The sequence of their products are copies of existing polynucleotide chains - which are called *templates*. These enzymes function by adding nucleotides onto the 32 hydroxyl group of the previous nucleotide in a DNA strand. As a consequence, all polymerases work in a 52 to 32 direction. In the active site of these enzymes, the incoming nucleoside triphosphate base-pairs to the template: this allows polymerases to accurately synthesize the complementary strand of their template. Polymerases are classified according to the type of template that they use.

In DNA replication, a DNA-dependent DNA polymerase makes a copy of a DNA sequence. Accuracy is vital in this process, so many of these polymerases have a proofreading activity. Here, the polymerase recognizes the occasional mistakes in the synthesis reaction by the lack of base pairing between the mismatched nucleotides. If a mismatch is detected, a 32 to 52 exonuclease activity is activated and the incorrect base removed. In most organisms, DNA polymerases function in a large complex called the replisome that contains multiple accessory subunits, such as the DNA clamp or helicases.

RNA-dependent DNA polymerases are a specialized class of polymerases that copy the sequence of an RNA strand into DNA. They include reverse transcriptase, which is a viral enzyme involved in the infection of cells by retroviruses, and telomerase, which is required for the replication of telomeres. Telomerase is an unusual polymerase because it contains its own RNA template as part of its structure.

Transcription is carried out by a DNA-dependent RNA polymerase that copies the sequence of a DNA strand into RNA. To begin transcribing a gene, the RNA polymerase binds to a sequence of DNA called a promoter and separates the DNA strands. It then copies the gene sequence into a messenger RNA transcript until it reaches a region of DNA called the terminator, where it halts and detaches from the DNA. As with human DNA-dependent DNA polymerases, RNA polymerase II, the enzyme that transcribes most of the genes in the human genome, operates as part of a large protein complex with multiple regulatory and accessory subunits.

Genetic Recombination

DNA helix usually does not interact with other segments of DNA, and in human cells the different chromosomes even occupy separate areas in the nucleus called "chromosome territories". This physical separation of different chromosomes is important for the ability of DNA to function as a stable repository for information, as one of the few times chromosomes interact is during chromosomal crossover when they recombine. Chromosomal crossover is when two DNA helices break, swap a section and then rejoin.

Recombination allows chromosomes to exchange genetic information and produces new combinations of genes, which increases the efficiency of natural selection and can be important in the rapid evolution of new proteins. Genetic recombination can also be involved in DNA repair, particularly in the cell's response to double-strand breaks.

The most common form of chromosomal crossover is homologous recombination, where the two chromosomes involved share very similar sequences. Non-homologous recombination can be damaging to cells, as it can produce chromosomal translocations and genetic abnormalities.

The recombination reaction is catalyzed by enzymes known as recombinases, such as RAD51. The first step in recombination is a double-stranded break either caused by an endonuclease or damage to the DNA. A series of steps catalyzed in part by the recombinase then leads to joining of the two helices by at least one Holliday junction, in which a segment of a single strand in each helix is annealed to the complementary strand in the other helix. The Holliday junction is a tetrahedral junction structure that can be moved along the pair of chromosomes, swapping one strand for another. The recombination reaction is then halted by cleavage of the junction and re-ligation of the released DNA.

Evolution

DNA contains the genetic information that allows all modern living things to function, grow and reproduce. However, it is unclear how long in the 4-billion-year history of life DNA has performed this function, as it has been proposed that the earliest forms of life may have used RNA as their genetic material. RNA may have acted as the central part of early cell metabolism as it can both transmit genetic information and carry out catalysis as part of ribozymes. This ancient RNA world where nucleic acid would have been used for both catalysis and genetics may have influenced the evolution of the current genetic code based on four nucleotide bases. This would occur, since the number of different bases in such an organism is a trade-off between a small number of bases increasing replication accuracy and a large number of bases increasing the catalytic efficiency of ribozymes.

However, there is no direct evidence of ancient genetic systems, as recovery of DNA from most fossils is impossible. This is because DNA will survive in the environment for less than one million years and slowly degrades into short fragments in solution. Claims for older DNA have been made, most notably a report of the isolation of a viable bacterium from a salt crystal 250 million years old, but these claims are controversial.

Uses in Technology

Genetic Engineering

Methods have been developed to purify DNA from organisms, such as phenol-chloroform extraction, and to manipulate it in the laboratory, such as restriction digests and the polymerase chain reaction. Modern biology and biochemistry make intensive use of these techniques in recombinant DNA technology. Recombinant DNA is a man-made DNA sequence that has been assembled from other DNA sequences. They can be transformed into organisms in the form of plasmids or in the appropriate format, by using a viral vector. The genetically modified organisms produced can be used to produce products such as recombinant proteins, used in medical research, or be grown in agriculture.

Forensics

Forensic scientists can use DNA in blood, semen, skin, saliva or hair found at a crime scene to identify a matching DNA of an individual, such as a perpetrator. This process is formally termed DNA profiling, but may also be called "genetic fingerprinting". In DNA profiling, the lengths of variable sections of repetitive DNA, such as short tandem repeats and minisatellites, are compared between people. This method is usually an extremely reliable technique for identifying a matching DNA. However, identification can be complicated if the scene is contaminated with DNA from several people. DNA profiling was developed in 1984 by British geneticist Sir Alec Jeffreys, and first used in forensic science to convict Colin Pitchfork in the 1988 Enderby murders case.

People convicted of certain types of crimes may be required to provide a sample of DNA for a database. This has helped investigators solve old cases where only a DNA sample was obtained from the scene. DNA profiling can also be used to identify victims of mass casualty incidents. On the other hand, many convicted people have been released from prison on the basis of DNA techniques, which were not available when a crime had originally been committed.

Bioinformatics

Bioinformatics involves the manipulation, searching, and data mining of biological data, and this includes DNA sequence data. The development of techniques to store and search DNA sequences have led to widely applied advances in computer science, especially string searching algorithms, machine learning and database theory. String searching or matching algorithms, which find an occurrence of a sequence of letters inside a larger sequence of letters, were developed to search for specific sequences of nucleotides. The DNA sequenced may be aligned with other DNA sequences to identify homologous sequences and locate the specific mutations that make them distinct. These techniques, especially multiple sequence alignment, are used in studying phylogenetic relationships and protein function. Data sets representing entire genomes' worth of DNA sequences, such as those produced by the Human Genome Project, are difficult to use without the annotations that identify the locations of genes and regulatory elements on each chromosome. Regions of DNA sequence that have the characteristic patterns associated with protein- or RNA-coding genes can be identified by gene finding algorithms, which allow researchers to predict the presence of particular gene products and their possible functions in an organism even before they have been isolated experimentally. Entire genomes may also be compared which can shed light on the evolutionary history of particular organism and permit the examination of complex evolutionary events.

DNA Nanotechnology

DNA nanotechnology uses the unique molecular recognition properties of DNA and other nucleic acids to create self-assembling branched DNA complexes with useful properties. DNA is thus used as a structural material rather than as a carrier of biological information. This has led to the creation of two-dimensional periodic lattices (both tile-based as well as using the "DNA origami" method) as well as three-dimensional structures in the shapes of polyhedra. Nanomechanical devices and algorithmic self-assembly have also been demonstrated, and these DNA structures have been used to template the arrangement of other molecules such as gold nanoparticles and streptavidin proteins.

History and Anthropology

Because DNA collects mutations over time, which are then inherited, it contains historical information, and, by comparing DNA sequences, geneticists can infer the evolutionary history of organisms, their phylogeny. This field of phylogenetics is a powerful tool in evolutionary biology. If DNA sequences within a species are compared, population geneticists can learn the history of particular populations. This can be used in studies ranging from ecological genetics to anthropology; For example, DNA evidence is being used to try to identify the Ten Lost Tribes of Israel. DNA has also been used to look at modern family relationships, such as establishing family relationships between the descendants of Sally Hemings and Thomas Jefferson. This usage is closely related to the use of DNA in criminal investigations detailed above. Indeed, some criminal investigations have been solved when DNA from crime scenes has matched relatives of the guilty individual.

History of DNA Research

DNA was first isolated by the Swiss physician Friedrich Miescher who, in 1869, discovered a microscopic substance in the pus of discarded surgical bandages. As it resided in the nuclei of cells, he called it "nuclein". In 1919, Phoebus Levene identified the base, sugar and phosphate nucleotide unit. Levene suggested that DNA consisted of a string of nucleotide units linked together through the phosphate groups. However, Levene thought the chain was short and the bases repeated in a fixed order. In 1937 William Astbury produced the first X-ray diffraction patterns that showed that DNA had a regular structure.

In 1928, Frederick Griffith discovered that traits of the "smooth" form of the *Pneumococcus* could be transferred to the "rough" form of the same bacteria by mixing killed "smooth" bacteria with the live "rough" form. This system provided the first clear suggestion that DNA carries genetic information—the Avery–MacLeod–McCarty experiment—when Oswald Avery, along with coworkers Colin MacLeod and Maclyn McCarty, identified DNA as the transforming principle in 1943. DNA's role in heredity was confirmed in 1952, when Alfred Hershey and Martha Chase in the Hershey–Chase experiment showed that DNA is the genetic material of the T2 phage. In 1953, James D. Watson and Francis Crick suggested what is now accepted as the first correct double-helix model of DNA structure in the journal *Nature*. Their double-helix, molecular model of DNA was then based on a single X-ray diffraction image (labeled as "Photo 51") taken by Rosalind Franklin and Raymond Gosling in May 1952, as well as the information that the DNA bases are paired — also obtained through private communications from Erwin Chargaff in the previous years. Chargaff's rules played a very important role in establishing double-helix configurations for B-DNA as well as A-DNA.

Experimental evidence supporting the Watson and Crick model were published in a series of five articles in the same issue of *Nature*. Of these, Franklin and Gosling's paper was the first publication of their own X-ray diffraction data and original analysis method that partially supported the Watson and Crick model; this issue also contained an article on DNA structure by Maurice Wilkins and two of his colleagues, whose analysis and *in vivo* B-DNA X-ray patterns also supported the presence *in vivo* of the double-helical DNA configurations as proposed by Crick and Watson for their double-helix molecular model of DNA in the previous two pages of *Nature*. In 1962, after Franklin's death, Watson, Crick, and Wilkins jointly received the Nobel Prize in Physiology or Medicine. However, Nobel rules of the time allowed only living recipients, but a vigorous debate continues on who should receive credit for the discovery.

In an influential presentation in 1957, Crick laid out the central dogma of molecular biology, which foretold the relationship between DNA, RNA, and proteins, and articulated the "adaptor hypothesis". Final confirmation of the replication mechanism that was implied by the double-helical structure followed in 1958 through the Meselson–Stahl experiment. Further work by Crick and coworkers showed that the genetic code was based on non-overlapping triplets of bases, called codons, allowing Har Gobind Khorana, Robert W. Holley and Marshall Warren Nirenberg to decipher the genetic code. These findings represent the birth of molecular biology.

Nucleotide

Nucleotides are molecules that, when joined together, make up the structural units of RNA and DNA. In addition, nucleotides play central roles in metabolism. In that capacity, they serve as sources of chemical energy (adenosine triphosphate and guanosine triphosphate), participate in cellular signaling (cyclic guanosine monophosphate and cyclic adenosine monophosphate), and are incorporated into important cofactors of enzymatic reactions (coenzyme A, flavin adenine dinucleotide, flavin mononucleotide, and nicotinamide adenine dinucleotide phosphate).

Nucleotide Structure

A nucleotide is composed of a nucleobase (nitrogenous base), a five-carbon sugar (either ribose or 2'-deoxyribose), and one to three phosphate groups.

Together, the nucleobase and sugar comprise a nucleoside. The phosphate groups form bonds with either the 2, 3, or 5-carbon of the sugar, with the 5-carbon site most common. Cyclic nucleotides form when the phosphate group is bound to two of the sugar's hydroxyl groups. Ribonucleotides are nucleotides where the sugar is ribose, and deoxyribonucleotides contain the sugar deoxyribose. Nucleotides can contain either a purine or a pyrimidine base. Nucleic acids are polymeric macromolecules made from nucleotide monomers. In DNA, the purine bases are adenine and guanine, while the pyrimidines are thymine and cytosine. RNA uses uracil in place of thymine. Adenine always pairs with thymine by 2 hydrogen bonds, while guanine pairs with cytosine through 3 hydrogen bonds, each due to their unique structures.

Synthesis

Nucleotides can be synthesized by a variety of means both in vitro and in vivo. In vivo, nucleotides can be synthesised de novo or recycled through salvage pathways. Nucleotides undergo breakdown such that useful parts can be reused in synthesis reactions to create new nucleotides. In vitro, protecting groups may be used during laboratory production of nucleotides. A purified nucleoside is protected to create a phosphoramidite, which can then be used to obtain analogues not found in nature and/or to synthesize an oligonucleotide.

Pyrimidine Ribonucleotides

Pyrimidine nucleotide synthesis starts with the formation of carbamoyl phosphate from glutamine and CO_2. The cyclisation reaction between carbamoyl phosphate reacts with aspartate, yielding orotate in subsequent steps. Orotate reacts with 5-phosphoribosyl á-diphosphate (PRPP), yielding orotidine monophosphate (OMP), which is decarboxylated to form uridine monophosphate (UMP). It is from UMP that other pyrimidine nucleotides are derived. UMP is phosphorylated to uridine triphosphate (UTP) via two sequential reactions with ATP. Cytidine monophosphate (CMP) is derived from conversion of UTP to cytidine triphosphate (CTP) with subsequent loss of two phosphates.

Purine Ribonucleotides

The atoms which are used to build the purine nucleotides come from a variety of sources:

The biosynthetic origins of purine ring atoms

N1 arises from the amine group of Asp

C2 and C8 originate from formate

N3 and N9 are contributed by the amide group of Gln

C4, C5 and N7 are derived from Gly

C6 comes from HCO_3^- (CO_2)

The de novo synthesis of purine nucleotides by which these precursors are incorporated into the purine ring proceeds by a 10-step pathway to the branch-point intermediate IMP, the nucleotide of the base hypoxanthine. AMP and GMP are subsequently synthesized from this intermediate via separate, two-step pathways. Thus, purine moieties are initially formed as part of the ribonucleotides rather than as free bases.

Six enzymes take part in IMP synthesis. Three of them are multifunctional:

- GART (reactions 2, 3, and 5)
- PAICS (reactions 6, and 7)
- ATIC (reactions 9, and 10)

Reaction 1. The pathway starts with the formation of PRPP. PRPS1 is the enzyme that activates R5P, which is formed primarily by the pentose phosphate pathway, to PRPP by reacting it with ATP. The reaction is unusual in that a pyrophosphoryl group is directly transferred from ATP to C1 of R5P and that the product has the á configuration about C1. This reaction is also shared with the pathways for the synthesis of the pyrimidine nucleotides, Trp, and His. As a result of being on (a) such (a) major metabolic crossroad and the use of energy, this reaction is highly regulated.

Reaction 2. In the first reaction unique to purine nucleotide biosynthesis, PPAT catalyzes the displacement of PRPP's pyrophosphate group (PP_i) by Gln's amide nitrogen. The reaction occurs with the inversion of configuration about ribose C1, thereby forming â-5-phosphorybosylamine (5-PRA) and establishing the anomeric form of the future nucleotide. This reaction, which is driven to completion by the subsequent hydrolysis of the released PP_i, is the pathway's flux-generating step and is therefore regulated, too.

Length Unit

Nucleotide (abbreviated nt) is a common length unit for single-stranded RNA, similar to how base pair is a length unit for double-stranded DNA.

RNA

Ribonucleic acid (RNA) is one of the three major macromolecules (along with DNA and proteins) that are essential for all known forms of life. Like DNA, RNA is made up of a long chain of components called nucleotides. Each nucleotide consists of a nucleobase (sometimes called a nitrogenous base), a ribose sugar, and a phosphate group. The sequence of nucleotides allows RNA to encode genetic information. For example, some viruses use RNA instead of DNA as their genetic material, and all organisms use messenger RNA (mRNA) to carry the genetic information that directs the synthesis of proteins. Like proteins, some RNA molecules play an active role in cells by catalyzing biological reactions, controlling gene expression, or sensing and communicating responses to cellular signals. One of these active processes is protein synthesis, a universal function whereby mRNA molecules direct the assembly of proteins on ribosomes.

This process uses transfer RNA (tRNA) molecules to deliver amino acids to the ribosome, where ribosomal RNA (rRNA) links amino acids together to form proteins.

The chemical structure of RNA is very similar to that of DNA, with two differences—(a) RNA contains the sugar ribose while DNA contains the slightly different sugar deoxyribose (a type of ribose that lacks one oxygen atom), and (b) RNA has the nucleobase uracil while DNA contains thymine (uracil and thymine have similar base-pairing properties).

Unlike DNA, most RNA molecules are single-stranded. Single-stranded RNA molecules adopt very complex three-dimensional structures, since they are not restricted to the repetitive double-helical form of double-stranded DNA. RNA is made within living cells by RNA polymerases, enzymes that act to copy a DNA or RNA template into a new RNA strand through processes known as transcription or RNA replication, respectively.

Comparison with DNA

RNA and DNA are both nucleic acids, but differ in three main ways. First, unlike DNA, which is, in general, double-stranded, RNA is a single-stranded molecule in many of its biological roles and has a much shorter chain of nucleotides. Second, while DNA contains *deoxyribose*, RNA contains *ribose* (in deoxyribose there is no hydroxyl group attached to the pentose ring in the 2' position).

These hydroxyl groups make RNA less stable than DNA because it is more prone to hydrolysis. Third, the complementary base to adenine is not thymine, as it is in DNA, but rather uracil, which is an unmethylated form of thymine.

Like DNA, most biologically active RNAs, including mRNA, tRNA, rRNA, snRNAs, and other non-coding RNAs, contain self-complementary sequences that allow parts of the RNA to fold and pair with itself to form double helices.

Structural analysis of these RNAs has revealed that they are highly structured. Unlike DNA, their structures do not consist of long double helices but rather collections of short helices packed together into structures akin to proteins.

In this fashion, RNAs can achieve chemical catalysis, like enzymes. For instance, determination of the structure of the ribosome—an enzyme that catalyzes peptide bond formation—revealed that its active site is composed entirely of RNA.

Structure

Each nucleotide in RNA contains a ribose sugar, with carbons numbered 1' through 5'. A base is attached to the 1' position, in general, adenine (A), cytosine (C), guanine (G), or uracil (U). Adenine and guanine are purines, cytosine, and uracil are pyrimidines. A phosphate group is attached to the 3' position of one ribose and the 5' position of the next. The phosphate groups have a negative charge each at physiological pH, making RNA a charged molecule (polyanion). The bases may form hydrogen bonds between cytosine and guanine, between adenine and uracil and between guanine and uracil. However, other interactions are possible, such as a group of adenine bases binding to each other in a bulge, or the GNRA tetraloop that has a guanine–adenine base-pair.

An important structural feature of RNA that distinguishes it from DNA is the presence of a hydroxyl group at the 2' position of the ribose sugar. The presence of this functional group causes the helix to adopt the A-form geometry rather than the B-form most commonly observed in DNA. This results in a very deep and narrow major groove and a shallow and wide minor groove. A second consequence of the presence of the 2'-hydroxyl group is that in conformationally flexible regions of an RNA molecule (that is, not involved in formation of a double helix), it can chemically attack the adjacent phosphodiester bond to cleave the backbone.

RNA is transcribed with only four bases (adenine, cytosine, guanine and uracil), but these bases and attached sugars can be modified in numerous ways as the RNAs mature. Pseudouridine (Ø), in which the linkage between uracil and ribose is changed from a C–N bond to a C–C bond, and ribothymidine (T) are found in various places (the most notable ones being in the TØC loop of tRNA). Another notable modified base is hypoxanthine, a deaminated adenine base whose nucleoside is called inosine (I). Inosine plays a key role in the wobble hypothesis of the genetic code.

There are nearly 100 other naturally occurring modified nucleosides, of which pseudouridine and nucleosides with 2'-O-methylribose are the most common. The specific roles of many of these modifications in RNA are not fully understood. However, it is notable that, in ribosomal RNA, many of the post-transcriptional modifications occur in highly functional regions, such as the peptidyl transferase center and the subunit interface, implying that they are important for normal function.

The functional form of single stranded RNA molecules, just like proteins, frequently requires a specific tertiary structure. The scaffold for this structure is provided by secondary structural elements that are hydrogen bonds within the molecule. This leads to several recognizable "domains" of secondary structure like hairpin loops, bulges, and internal loops. Since RNA is charged, metal ions such as Mg^{2+} are needed to stabilise many secondary and tertiary structures.

Synthesis

Synthesis of RNA is usually catalyzed by an enzyme—RNA polymerase—using DNA as a template, a process known as transcription. Initiation of transcription begins with the binding of the enzyme to a promoter sequence in the DNA (usually found "upstream" of a gene). The DNA double helix is unwound by the helicase activity of the enzyme. The enzyme then progresses along the template strand in the 3' to 5' direction, synthesizing a complementary RNA molecule with elongation occurring in the 5' to 3' direction. The DNA sequence also dictates where termination of RNA synthesis will occur. RNAs are often modified by enzymes after transcription. For example, a poly(A) tail and a 5' cap are added to eukaryotic pre-mRNA and introns are removed by the spliceosome.

There are also a number of RNA-dependent RNA polymerases that use RNA as their template for synthesis of a new strand of RNA. For instance, a number of RNA viruses (such as poliovirus) use this type of enzyme to replicate their genetic material. Also, RNA-dependent RNA polymerase is part of the RNA interference pathway in many organisms.

Types of RNA

Overview

Messenger RNA (mRNA) is the RNA that carries information from DNA to the ribosome, the sites of protein synthesis (translation) in the cell. The coding sequence of the mRNA determines the amino acid sequence in the protein that is produced. Many RNAs do not code for protein however (about 97% of the transcriptional output is non-protein-coding in eukaryotes).

These so-called non-coding RNAs ("ncRNA") can be encoded by their own genes (RNA genes), but can also derive from mRNA introns. The most prominent examples of non-coding RNAs are transfer RNA (tRNA) and ribosomal RNA (rRNA), both of which are involved in the process of translation. There are also non-coding RNAs involved in gene regulation, RNA processing and other roles. Certain RNAs are able to catalyse chemical reactions such as cutting and ligating other RNA molecules, and the catalysis of peptide bond formation in the ribosome; these are known as ribozymes.

In Translation

Messenger RNA (mRNA) carries information about a protein sequence to the ribosomes, the protein synthesis factories in the cell. It is coded so that every three nucleotides (a codon) correspond to one amino acid. In eukaryotic cells, once precursor mRNA (pre-mRNA) has been transcribed from DNA, it is processed to mature mRNA. This removes its introns—non-coding sections of the pre-mRNA. The mRNA is then exported from the nucleus to the cytoplasm, where it is bound to ribosomes and translated into its corresponding protein form with the help of tRNA. In prokaryotic cells, which do not have nucleus and cytoplasm compartments, mRNA can bind to ribosomes while it is being transcribed from DNA. After a certain amount of time the message degrades into its component nucleotides with the assistance of ribonucleases.

Transfer RNA (tRNA) is a small RNA chain of about 80 nucleotides that transfers a specific amino acid to a growing polypeptide chain at the ribosomal site of protein synthesis during translation. It has sites for amino acid attachment and an anticodon region for codon recognition that binds to a specific sequence on the messenger RNA chain through hydrogen bonding. Ribosomal RNA (rRNA) is the catalytic component of the ribosomes. Eukaryotic ribosomes contain four different rRNA molecules: 18S, 5.8S, 28S and 5S rRNA. Three of the rRNA molecules are synthesized in the nucleolus, and one is synthesized elsewhere. In the cytoplasm, ribosomal RNA and protein combine to form a nucleoprotein called a ribosome.

The ribosome binds mRNA and carries out protein synthesis. Several ribosomes may be attached to a single mRNA at any time. rRNA is extremely abundant and makes up 80% of the 10 mg/ml RNA found in a typical eukaryotic cytoplasm. Transfer-messenger RNA (tmRNA) is found in many bacteria and plastids. It tags proteins encoded by mRNAs that lack stop codons for degradation and prevents the ribosome from stalling.

Regulatory RNAs

Several types of RNA can downregulate gene expression by being complementary to a part of an mRNA or a gene's DNA. MicroRNAs (miRNA; 21-22 nt) are found in eukaryotes and act through RNA interference (RNAi), where an effector complex of miRNA and enzymes can break down mRNA to which the miRNA is complementary, block the mRNA from being translated, or accelerate its degradation. While small interfering RNAs (siRNA; 20-25 nt) are often produced by breakdown of viral RNA, there are also endogenous sources of siRNAs.

siRNAs act through RNA interference in a fashion similar to miRNAs. Some miRNAs and siRNAs can cause genes they target to be methylated, thereby decreasing or increasing transcription of those genes. Animals have Piwi-interacting RNAs (piRNA; 29-30 nt) which are active in germline cells and are thought to be a defense against transposons and play a role in gametogenesis. Many prokaryotes have CRISPR RNAs, a regulatory system similar to RNA interference. Antisense RNAs are widespread; most downregulate a gene, but a few are activators of transcription. One way antisense RNA can act is by binding to an mRNA, forming double-stranded RNA that is enzymatically degraded. There are many long noncoding RNAs that regulate genes in eukaryotes, one such RNA is Xist, which coats one X chromosome in female mammals and inactivates it. An mRNA may contain regulatory elements itself, such as riboswitches, in the 5' untranslated region or 3' untranslated region; these cis-regulatory elements regulate the activity of that mRNA. The untranslated regions can also contain elements that regulate other genes.

In RNA Processing

Many RNAs are involved in modifying other RNAs. Introns are spliced out of pre-mRNA by spliceosomes, which contain several small nuclear RNAs (snRNA), or the introns can be ribozymes that are spliced by themselves. RNA can also be altered by having its nucleotides modified to other nucleotides than A, C, G and U. In eukaryotes, modifications of RNA nucleotides are generally directed by small nucleolar RNAs (snoRNA; 60-300 nt), found in the nucleolus and cajal bodies. snoRNAs associate with enzymes and guide them to a spot on an RNA by basepairing to that RNA. These enzymes then perform the nucleotide modification. rRNAs and tRNAs are extensively modified, but snRNAs and mRNAs can also be the target of base modification.

RNA Genomes

Like DNA, RNA can carry genetic information. RNA viruses have genomes composed of RNA, and a variety of proteins encoded by that genome. The viral genome is replicated by some of those proteins, while other proteins protect the genome as the virus particle moves to a new host cell. Viroids are another group of pathogens, but they consist only of RNA, do not encode any protein and are replicated by a host plant cell's polymerase.

In Reverse Transcription

Reverse transcribing viruses replicate their genomes by reverse transcribing DNA copies from their RNA; these DNA copies are then transcribed to new RNA. Retrotransposons also spread by copying DNA and RNA from one another, and telomerase contains an RNA that is used as template for building the ends of eukaryotic chromosomes.

Double-Stranded RNA

Double-stranded RNA (dsRNA) is RNA with two complementary strands, similar to the DNA found in all cells. dsRNA forms the genetic material of some viruses (double-stranded RNA viruses). Double-stranded RNA such as viral RNA or siRNA can trigger RNA interference in eukaryotes, as well as interferon response in vertebrates.

Key Discoveries in RNA Biology

Research on RNA has led to many important biological discoveries and numerous Nobel Prizes. Nucleic acids were discovered in 1868 by Friedrich Miescher, who called the material 'nuclein' since it was found in the nucleus. It was later discovered that prokaryotic cells, which do not have a nucleus, also contain nucleic acids. The role of RNA in protein synthesis was suspected already in 1939. Severo Ochoa won the 1959 Nobel Prize in Medicine (shared with Arthur Kornberg) after he discovered an enzyme that can synthesize RNA in the laboratory. Ironically, the enzyme discovered by Ochoa (polynucleotide phosphorylase) was later shown to be responsible for RNA degradation, not RNA synthesis.

The sequence of the 77 nucleotides of a yeast tRNA was found by Robert W. Holley in 1965, winning Holley the 1968 Nobel Prize in Medicine (shared with Har Gobind Khorana and Marshall Nirenberg). In 1967, Carl Woese hypothesized that RNA might be catalytic and suggested that the earliest forms of life (self-replicating molecules) could have relied on RNA both to carry genetic information and to catalyze biochemical reactions—an RNA world. During the early 1970s, retroviruses and reverse transcriptase were discovered, showing for the first time that enzymes could copy RNA into DNA (the opposite of the usual route for transmission of genetic information). For this work, David Baltimore, Renato Dulbecco and Howard Temin were awarded a Nobel Prize in 1975. In 1976, Walter Fiers and his team determined the first complete nucleotide sequence of an RNA virus genome, that of bacteriophage MS2.

In 1977, introns and RNA splicing were discovered in both mammalian viruses and in cellular genes, resulting in a 1993 Nobel to Philip Sharp and Richard Roberts. Catalytic RNA molecules (ribozymes) were discovered in the early 1980s, leading to a 1989 Nobel award to Thomas Cech and Sidney Altman. In 1990 it was found in petunia that introduced genes can silence similar genes of the plant's own, now known to be a result of RNA interference. At about the same time, 22 nt long RNAs, now called microRNAs, were found to have a role in the development of *C. elegans*. Studies on RNA interference gleaned a Nobel Prize for Andrew Fire and Craig Mello in 2006, and another Nobel was awarded for studies on transcription of RNA to Roger Kornberg in the same year. The discovery of gene regulatory RNAs has led to attempts to develop drugs made of RNA, such as siRNA, to silence genes.

Genome

In modern molecular biology and genetics, the genome is the entirety of an organism's hereditary information. It is encoded either in DNA or, for many types of virus, in RNA. The genome includes both the genes and the non-coding sequences of the DNA/RNA.

The term was adapted in 1920 by Hans Winkler, Professor of Botany at the University of Hamburg, Germany. In Greek, the word *genome* (ãßíïìáé) means "I become, I am born, to come into being". The Oxford English Dictionary suggests the name to be a blend of the words *gene* and *chromosome*. A few related *-ome* words already existed, such as *biome* and *rhizome*, forming a vocabulary into which *genome* fits systematically.

Some organisms have multiple copies of chromosomes, diploid, triploid, tetraploid and so on. In classical genetics, in a sexually reproducing organism (typically eukarya) the gamete has half of the number of chromosome of the somatic cell and the genome is a full set of chromosomes in a gamete. In haploid organisms, including cells of bacteria, archaea, and in organelles including mitochondria and chloroplasts, or viruses, that similarly contain genes, the single or set of circular and/or linear chains of DNA (or RNA for some viruses), likewise constitute the *genome*. The term genome can be applied specifically to mean that stored on a complete set of *nuclear DNA* (i.e., the "nuclear genome") but can also be applied to that stored within organelles that contain their own DNA, as with the "mitochondrial genome" or the "chloroplast genome". Additionally, the genome can comprise nonchromosomal genetic elements such as viruses, plasmids, and transposable elements.

When people say that the genome of a sexually reproducing species has been "sequenced", typically they are referring to a determination of the sequences of one set of autosomes and one of each type of sex chromosome, which together represent both of the possible sexes. Even in species that exist in only one sex, what is described as "a genome sequence" may be a composite read from the chromosomes of various individuals. In general use, the phrase "genetic makeup" is sometimes used conversationally to mean the genome of a particular individual or organism. The study of the global properties of genomes of related organisms is usually referred to as genomics, which distinguishes it from genetics which generally studies the properties of single genes or groups of genes. Both the number of base pairs and the number of genes vary widely from one species to another, and there is only a rough correlation between the two (an observation known as the C-value paradox). At present, the highest known number of genes is around 60,000, for the protozoan causing trichomoniasis, almost three times as many as in the human genome.

An analogy to the human genome stored on DNA is that of instructions stored in a book:

- The book (genome) would contain 23 chapters (chromosomes);
- each chapter contains 48 to 250 million letters (A,C,G,T) without spaces;
- Hence, the book contains over 3.2 billion letters total;
- The book fits into a cell nucleus the size of a pinpoint;

- At least one copy of the book (all 23 chapters) is contained in most cells of our body. The red blood cells do not have nuclei, and therefore do not have a copy of the genome. Sexual reproduction cells (Ova and spermatozoids) have only half of the genome.

Most biological entities that are more complex than a virus sometimes or always carry additional genetic material besides that which resides in their chromosomes. In some contexts, such as sequencing the genome of a pathogenic microbe, "genome" is meant to include information stored on this auxiliary material, which is carried in plasmids. In such circumstances then, "genome" describes all of the genes and information on non-coding DNA that have the potential to be present. In eukaryotes such as plants, protozoa and animals, however, "genome" carries the typical connotation of only information on chromosomal DNA. So although these organisms contain chloroplasts and/or mitochondria that have their own DNA, the genetic information contained by DNA within these organelles is not considered part of the genome. In fact, mitochondria are sometimes said to have their own genome often referred to as the "mitochondrial genome". The DNA found within the chloroplast may be referred to as the "plastome".

Genomes and Genetic Variation

Note that a genome does not capture the genetic diversity or the genetic polymorphism of a species. For example, the human genome sequence in principle could be determined from just half the information on the DNA of one cell from one individual. To learn what variations in genetic information underlie particular traits or diseases requires comparisons across individuals. This point explains the common usage of "genome" (which parallels a common usage of "gene") to refer not to the information in any particular DNA sequence, but to a whole family of sequences that share a biological context.

Although this concept may seem counter intuitive, it is the same concept that says there is no particular shape that is the shape of a cheetah. Cheetahs vary, and so do the sequences of their genomes. Yet both the individual animals and their sequences share commonalities, so one can learn something about cheetahs and "cheetah-ness" from a single example of either.

Sequencing and Mapping

The Human Genome Project was organized to map and to sequence the human genome. Other genome projects include mouse, rice, the plant *Arabidopsis thaliana*, the puffer fish, and bacteria like E. coli. In 1976, Walter Fiers at the University of Ghent (Belgium) was the first to establish the complete nucleotide sequence of a viral RNA-genome (bacteriophage MS2). The first DNA-genome project to be completed was the Phage Ö-X174, with only 5386 base pairs, which was sequenced by Fred Sanger in 1977. The first bacterial genome to be completed was that of Haemophilus influenzae, completed by a team at The Institute for Genomic Research in 1995. A few months later, the first eukaryotic genome was completed, with the 16 chromosomes of budding yeast Saccharomyces cerevisiae being released as a result of a European led effort initiated in the mid-80's. The development of new technologies has dramatically decreased the difficulty and cost of sequencing, and the number of complete genome sequences is rising rapidly. Among many genome database sites, the one maintained by the US National Institutes of Health is inclusive.

These new technologies open up the prospect of personal genome sequencing as an important diagnostic tool. A major step toward that goal was the completion of the decipherment of the full genome of DNA pioneer James D. Watson in 2007. Whereas a genome sequence lists the order of every DNA base in a genome, a genome map identifies the landmarks. A genome map is less detailed than a genome sequence and aids in navigating around the genome. A fundamental step in the Human genome project was the release of a detailed genomic map by Jean Weissenbach and his team at the Genoscope in Paris

Since genomes and their organisms are very complex, one research strategy is to reduce the number of genes in a genome to the bare minimum and still have the organism in question survive. There is experimental work being done on minimal genomes for single cell organisms as well as minimal genomes for multicellular organisms. The work is both *in vivo* and *in silico.*

Genomes are more than the sum of an organism's genes and have traits that may be measured and studied without reference to the details of any particular genes and their products. Researchers compare traits such as *chromosome number* (karyotype), genome size, gene order, codon usage bias, and GC-content to determine what mechanisms could have produced the great variety of genomes that exist today (for recent overviews, see Brown 2002; Saccone and Pesole 2003; Benfey and Protopapas 2004; Gibson and Muse 2004; Reese 2004; Gregory 2005).

Duplications play a major role in shaping the genome. Duplications may range from extension of short tandem repeats, to duplication of a cluster of genes, and all the way to duplications of entire chromosomes or even entire genomes. Such duplications are probably fundamental to the creation of genetic novelty. Horizontal gene transfer is invoked to explain how there is often extreme similarity between small portions of the genomes of two organisms that are otherwise very distantly related. Horizontal gene transfer seems to be common among many microbes. Also, eukaryotic cells seem to have experienced a transfer of some genetic material from their chloroplast and mitochondrial genomes to their nuclear chromosomes.

Heredity

Heredity is the passing of traits to offspring (from its parent or ancestors). This is the process by which an offspring cell or organism acquires or becomes predisposed to the characteristics of its parent cell or organism. Through heredity, variations exhibited by individuals can accumulate and cause some species to evolve. The study of heredity in biology is called genetics, which includes the field of epigenetics.

In humans, eye color is an inherited characteristic and an individual might inherit the "brown-eye trait" from one of the parents. Inherited traits are controlled by genes and the complete set of genes within an organism's genome is called its genotype.

The complete set of observable traits that make up the structure and behaviour of an organism is called its phenotype. These traits come from the interaction of its genotype with the environment. As a result, many aspects of an organism's phenotype are not inherited. For example, suntanned skin comes from the interaction between a person's genotype and sunlight; thus, suntans are not passed on to people's children. However, some people tan more easily than others, due to differences in their genotype; a striking example are people with the inherited trait of albinism, who do not tan at all and are very sensitive to sunburn.

Heritable traits are known to be passed from one generation to the next via DNA, a molecule that encodes genetic information. DNA is a long polymer composed of four types of bases. The sequence of bases along a particular DNA molecule specify the genetic information, in a manner similar to a sequence of letters spelling out a sentence. Before a cell divides, the DNA is copied, so that each of the resulting two cells will inherit the DNA sequence. Portions of a DNA molecule that specify a single functional unit are called genes; different genes have different sequences of bases. Within cells, the long strands of DNA form condensed structures called chromosomes. The specific location of a DNA sequence within a chromosome is known as a locus. If the DNA sequence at a locus varies between individuals, the different forms of this sequence are called alleles. DNA sequences can change through mutations, producing new alleles. If a mutation occurs within a gene, the new allele may affect the trait that the gene controls, altering the phenotype of the organism.

However, while this simple correspondence between an allele and a trait works in some cases, most traits are more complex and are controlled by multiple interacting genes within and among organisms. Developmental biologists suggest that complex interactions in genetic networks and communication among cells can lead to heritable variations that may underlay some of the mechanics in developmental plasticity and canalization.

Recent findings have confirmed important examples of heritable changes that cannot be explained by direct agency of the DNA molecule. These phenomena are classed as epigenetic inheritance systems that are causally or independently evolving over genes. Research into modes and mechanisms of epigenetic inheritance is still in its scientific infancy, however, this area of research has attracted much recent activity as it broadens the scope of heritability and evolutionary biology in general. DNA methylation marking chromatin, self-sustaining metabolic loops, gene silencing by RNA interference, and the three dimensional conformation of proteins (such as prions) are areas where epigeneti inheritance ystems have been dicovered at the organismic level. Heritability may also occur at even larger scales. For example, ecological inheritance through the process of niche construction is defined by the regular and repeated activities of organisms in their environment.

This generates a legacy of effect that modifies and feeds back into the selection regime of subsequent generations. Descendants inherit genes plus environmental characteristics generated by the ecological actions of ancestors. Other examples of heritability in evolution that are not under the direct control of genes include the inheritance of cultural traits, group heritability, and symbiogenesis. These examples of heritability that operate above the gene are covered broadly under the title of multilevel or hierarchical selection, which has been a subject of intense debate in the history of evolutionary science.

Relation to Theory of Evolution

When Charles Darwin proposed his theory of evolution in 1859, one of its major problems was the lack of an underlying mechanism for heredity. Darwin believed in a mix of blending inheritance and the inheritance of acquired traits (pangenesis). Blending inheritance would lead to uniformity across populations in only a few generations and thus would remove variation from a population on which natural selection could act. This led to Darwin adopting some Lamarckian ideas in later editions of *On the Origin of Species* and his later biological works. Darwin's primary approach to heredity was to outline how it appeared to work (noticing that traits could be inherited which were not expressed explicitly in the parent at the time of reproduction, that certain traits could be sex-linked, etc.) rather than suggesting mechanisms.

Darwin's initial model of herediy was adopted by, and then havily modified by, his cousin Francis Galton, who laid the framework for the biometric school of heredity. Galton rejected the aspects of Darwin's pangeneis model which relied on acquired traits.

The inheritance of acquired traits was shown to have little basis in the 1880s when August Weismann cut the tails off many generations of mice and found that their offspring continued to develop tails.

The ancients had a variety of ideas about heredity: Theophrastus proposed that male flowers caused female flowers to ripen; Hippocrates speculated that "seeds" were produced by various body parts and transmitted to offspring at the time of conception, and Aristotle thought that male and female semen mixed at conception. Aeschylus, in 458 BC, proposed the male as the parent, with the female as a "nurse for the young life sown within her."

Various hereditary mechanisms were envisaged without being properly tested or quantified. These included blending inheritance and the inheritance of acquired traits. Nevertheless, people were able to develop domestic breeds of animals as well as crops through artificial selection. The inheritance of acquired traits also formed a part of early Lamarckian ideas on evolution.

In the 9th century AD, the Afro-Arab writer Al-Jahiz considered the effects of the environment on the likelihood of an animal to survive, and first described the struggle for existence. His ideas on the struggle for existence in the *Book of Animals* have been summarized as follows:

Animals engage in a struggle for existence; for resources, to avoid being eaten and to breed. Environmental factors influence organisms to develop new characteristics to ensure survival, thus transforming into new species. Animals that survive to breed can pass on their successful characteristics to offspring.

In 1000 AD, the Arab physician, Abu al-Qasim al-Zahrawi (known as Albucasis in the West), wrote the first clear description of haemophilia, a hereditary genetic disorder, in his *Al-Tasrif*. In this work, he wrote of an Andalusian family whose males died of bleeding after minor injuries.

During the 18th century, Dutch microscopist Antonie van Leeuwenhoek (1632–1723) discovered "animalcules" in the sperm of humans and other animals. Some scientists speculated they saw a "little man" (homunculus) inside each sperm. These scientists formed a school of thought known as the "spermists." They contended the only contributions of the female to the next generation were the womb in which the homunculus grew, and prenatal influences of the womb. An opposing school of thought, the ovists, believed that the future human was in the egg, and that perm merely stimulated the growth of the egg. Ovists thought women carried eggs containing boy and girl children, and that the gender of the offspring was determined well before conception.

Gregor Mendel: Father of Modern Genetics

The idea of particulate inheritance of genes can be attributed to the Moravian monk Gregor Mendel who published his work on pea plants in 1865. However, his work was not widely known and was rediscovered in 1901. It was initially assumed the Mendelian inheritance only accounted for large (qualitative) differences, such as those seen by Mendel in his pea plants—and the idea of additive effect of (quantitative) genes was not realized until R.A. Fisher's (1918) paper, "The Correlation Between Relatives on the Supposition of Mendelian Inheritance."

Modern Development of Genetics and Heredity

In the 1930s, work by Fisher and others resulted in a combination of Mendelian and biometric schools into the modern evolutionary synthesis. The modern synthesis bridged the gap between experimental geneticists and naturalists; and between both and palaeontologists, stating that:

1. All evolutionary phenomena can be explained in a way consistent with known genetic mechanisms and the observational evidence of naturalists.
2. Evolution is gradual: small genetic changes, recombination ordered by natural selection. Discontinuities amongst species (or other taxa) are explained as originating gradually through geographical separation and extinction (not saltation).
3. Selection is overwhelmingly the main mechanism of change; even slight advantages are important when continued. The object of selection is the phenotype in its surrounding environment. The role of genetic drift is equivocal; though strongly supported initially by Dobzhansky, it was downgraded later as results from ecological genetics were obtained.
4. The primacy of population thinking: the genetic diversity carried in natural populations is a key factor in evolution. The strength of natural selection in the wild was greater than expected; the effect of ecological factors such as niche occupation and the significance of barriers to gene flow are all important.
5. In palaeontology, the ability to explain historical observations by extrapolation from micro to macro-evolution is proposed. Historical contingency means explanations at different levels may exist. Gradualism does not mean constant rate of change.

The idea that speciation occurs after populations are reproductively isolated has been much debated. In plants, polyploidy must be included in any view of speciation. Formulations such as 'evolution consists primarily of changes in the frequencies of alleles between one generation and another' were proposed rather later. The traditional view is that developmental biology ('evo-devo') played little part in the synthesis, but an account of Gavin de Beer's work by Stephen Jay Gould suggests he may be an exception.

Almost all aspects of the synthesis have been challenged at times, with varying degrees of success. There is no doubt, however, that the synthesis was a great landmark in evolutionary biology. It cleared up many confusions, and was directly responsible for stimulating a great deal of research in the post-World War II era.

Trofim Lysenko however caused a backlash of what is now called Lysenkoism in the Soviet Union when he emphasised Lamarckian ideas on the inheritance of acquired traits. This movement affected agricultural research and led to food shortages in the 1960s and seriously affected the USSR.

Types of Heredity

Dominant and Recessive

An allele is said to be dominant if it is always expressed in the appearance of an organism (phenotype). For example, in peas the allele for gree pods, G, is dominant to that for yellow pods, g. Since the allele for green pods is dominant, pea plants with the pair of alleles GG (homozygote) or Gg (heterozygote) will have green pods. The allele for yellow pods is recessive. The effects of this allele are only seen when it is present in both chromosomes, gg (homozygote).

The description of a mode of biological inheritance consists of three main categories:

1. Number of involved loci
    - Monogenetic (also called "simple") – one locus
    - Oligogenetic – few loci
    - Polygenetic – many loci
2. Involved chromosomes
    - Autosomal – loci are not situated on a sex chromosome
    - Gonosomal – loci are situated on a sex chromosome
        - X-chromosomal – loci are situated on the X-chromosome (the more common case)
        - Y-chromosomal – loci are situated on the Y-chromosome
    - Mitochondrial – loci are situated on the mitochondrial DNA
3. Correlation genotype–phenotype
    - Dominant
    - Intermediate (also called "codominant")
    - Recessive

These three categories are part of every exact description of a mode of inheritance in the above order. Additionally, more specifications may be added as follows:

4. Coincidental and environmental interactions
    - Penetrance
        - o Complete
        - o Incomplete (percentual number)
    - Expressivity
        - o Invariable
        - o Variable
    - Heritability (in polygenetic and sometimes also in oligogenetic modes of inheritance)
    - Maternal or paternal imprinting phenomena (also see epigenetics)
5. Sex-linked interactions
    - Sex-linked inheritance (gonosomal loci)
    - Sex-limited phenotype expression (e.g., cryptorchism)
    - Inheritance through the maternal line (in case of mitochondrial DNA loci)
    - Inheritance through the paternal line (in case of Y-chromosomal loci)
6. Locus–locus interactions
    - Epistasis with other loci (e.g., overdominance)
    - Gene coupling with other loci (also see crossing over)
    - Homozygotous lethal factors
    - Semi-lethal factors

Determination and description of a mode of inheritance is primarily achieved through statistical analysis of pedigree data. In case the involved loci are known, methods of molecular genetics can also be employed.

Mutation

In molecular biology and genetics, mutations are changes in a genomic sequence: the DNA sequence of a cell's genome or the DNA or RNA sequence of a virus. They can be defined as sudden and spontaneous changes in the cell. Mutations are caused by radiation, viruses, transposons and mutagenic chemicals, as well as errors that occur during meiosis or DNA replication. They can also be induced by the organism itself, by cellular processes such as hypermutation. Mutation can result in several different types of change in sequences;(DNA) these can either have no effect, alter the product of a gene, or prevent the gene from functioning properly or completely.

Studies in the fly *Drosophila melanogaster* suggest that if a mutation changes a protein produced by a gene, this will probably be harmful, with about 70 percent of these mutations having damaging effects, and the remainder being either neutral or weakly beneficial. Due to the damaging effects that mutations can have on genes, organisms have mechanisms such as DNA repair to remove mutations. Therefore, the optimal mutation rate for a species is a trade-off between costs of a high mutation rate, such as deleterious mutations, and the metabolic costs of maintaining systems to reduce the mutation rate, such as DNA repair enzymes. Viruses that use RNA as their genetic material have rapid mutation rates, which can be an advantage since these viruses will evolve constantly and rapidly, and thus evade the defensive responses of e.g. the human immune system.

Mutations can involve large sections of DNA becoming duplicated, usually through genetic recombination. These duplications are a major source of raw material for evolving new genes, with tens to hundreds of genes duplicated in animal genomes every million years. Most genes belong to larger families of genes of shared ancestry. Novel genes are produced by several methods, commonly through the duplication and mutation of an ancestral gene, or by recombining parts of different genes to form new combinations with new functions.

Here, domains act as modules, each with a particular and independent function, that can be mixed together to produce genes encoding new proteins with novel properties. For example, the human eye uses four genes to make structures that sense light: three for color vision and one for night vision; all four arose from a single ancestral gene. Another advantage of duplicating a gene (or even an entire genome) is that this increases redundancy; this allows one gene in the pair to acquire a new function while the other copy performs the original function. Other types of mutation occasionally create new genes from previously noncoding DNA.

Changes in chromosome number may involve even larger mutations, where segments of the DNA within chromosomes break and then rearrange. For example, two chromosomes in the *Homo* genus fused to produce human chromosome 2; this fusion did not occur in the lineage of the other apes, and they retain these separate chromosomes. In evolution, the most important role of such chromosomal rearrangements may be to accelerate the divergence of a population into new species by making populations less likely to interbreed, and thereby preserving genetic differences between these populations.

Sequences of DNA that can move about the genome, such as transposons, make up a major fraction of the genetic material of plants and animals, and may have been important in the evolution of genomes. For example, more than a million copies of the Alu equence are present in the human genome, and these sequences have now been recruited to perform functions such as regulating gene expression. Another effect of these mobile DNA sequences is that when they move within a genome, they can mutate or delete existing genes and thereby produce genetic diversity.

In multicellular organisms with dedicated reproductive cells, mutations can be subdivided into germ line mutations, which can be passed on to descendants through their reproductive cells, and somatic mutations (also called acquired mutations), which involve cells outside the dedicated reproductive group and which are not usually transmitted to descendants. If the organism can reproduce asexually through mechanisms such as cuttings or budding the distinction can become blurred. For example, plants can sometimes transmit somatic mutations to their descendants asexually or sexually where flower buds develop in somatically mutated parts of plants. A new mutation that was not inherited from either parent is called a *de novo* mutation. The source of the mutation is unrelated to the consequence, although the consequences are related to which cells were mutated.

Nonlethal mutations accumulate within the gene pool and increase the amount of genetic variation. The abundance of some genetic changes within the gene pool can be reduced by natural selection, while other "more favourable" mutations may accumulate and result in adaptive changes. For example, a butterfly may produce offspring with new mutations. The majority of these mutations will have no effect; but one might change the color of one of the butterfly's offspring, making it harder (or easier) for predators to see. If this color change is advantageous, the chance of this butterfly surviving and producing its own offspring are a little better, and over time the number of butterflies with this mutation may form a larger percentage of the population.

Neutral mutations are defined as mutations whose effects do not influence the fitness of an individual. These can accumulate over time due to genetic drift. It is believed that the overwhelming majority of mutations have no significant effect on an organism's fitness. Also, DNA repair mechanisms are able to mend most changes before they become permanent mutations, and many organisms have mechanisms for eliminating otherwise permanently mutated somatic cells. Mutation is generally accepted by biologists as the mechanism by which natural selection acts, generating advantageous new traits that survive and multiply in offspring as well as disadvantageous traits, in less fit offspring, that tend to die out.

Causes

Two classes of mutations are spontaneous mutations (molecular decay) and induced mutations caused by mutagens.

Spontaneous mutations on the molecular level can be caused by:

- Tautomerism – A base is changed by the repositioning of a hydrogen atom, altering the hydrogen bonding pattern of that base resulting in incorrect base pairing during replication.
- Depurination – Loss of a purine base (A or G) to form an apurinic site (AP site).
- Deamination – Hydrolysis changes a normal base to an atypical base containing a keto group in place of the original amine group. Examples include C '! U and A '! HX (hypoxanthine), which can be corrected by DNA repair mechanisms; and 5MeC (5-methylcytosine) '! T, which is less likely to be detected as a mutation because thymine is a normal DNA base.
- Slipped strand mispairing - Denaturation of the new strand from the template during replication, followed by renaturation in a different spot ("slipping"). This can lead to insertions or deletions.

Induced mutations on the molecular level can be caused by:

- Chemicals
  - o Hydroxylamine NH_2OH
  - o Base analogs (e.g. BrdU)
  - o Alkylating agents (e.g. *N*-ethyl-*N*-nitrosourea) These agents can mutate both replicating and non-replicating DNA. In contrast, a base analog can only mutate the DNA when the analog is incorporated in replicating the DNA. Each of these classes of chemical mutagens has certain effects that then lead to transitions, transversions, or deletions.
  - o Agents that form DNA adducts (e.g. ochratoxin A metabolites)
  - o DNA intercalating agents (e.g. ethidium bromide)
  - o DNA crosslinkers
  - o Oxidative damage
  - o Nitrous acid converts amine groups on A and C to diazo groups, altering their hydrogen bonding patterns which leads to incorrect base pairing during replication.
- Radiation
  - o Ultraviolet radiation (nonionizing radiation). Two nucleotide bases in DNA – cytosine and thymine – are most vulnerable to radiation that can change their properties. UV light can induce adjacent pyrimidine bases in a DNA strand to become covalently joined as a pyrimidine dimer. UV radiation, particularly longer-wave UVA, can also cause oxidative damage to DNA.

- o Ionizing radiation
- o Radioactive decay, such as ^{14}C in DNA
- Viral infections

DNA has so-called hotspots, where mutations occur up to 100 times more frequently than the normal mutation rate. A hotspot can be at an unusual base, e.g., 5-methylcytosine.

Mutation rates also vary across species. Evolutionary biologists have theorized that higher mutation rates are beneficial in some situations, because they allow organisms to evolve and therefore adapt more quickly to their environments. For example, repeated exposure of bacteria to antibiotics, and selection of resistant mutants, can result in the selection of bacteria that have a much higher mutation rate than the original population (mutator strains).

Classification of Mutation Types

By Effect on Structure

The sequence of a gene can be altered in a number of ways. Gene mutations have varying effects on health depending on where they occur and whether they alter the function of essential proteins. Mutations in the structure of genes can be classified as:

- Small-scale mutations, such as those affecting a small gene in one or a few nucleotides, including:
  - o Point mutations, often caused by chemicals or malfunction of DNA replication, exchange a single nucleotide for another. These changes are classified as transitions or transversions. Most common is the transition that exchanges a purine for a purine (A "! G) or a pyrimidine for a pyrimidine, (C "! T). A transition can be caused by nitrous acid, base mis-pairing, or mutagenic base analogs such as 5-bromo-2-deoxyuridine (BrdU). Less common is a transversion, which exchanges a purine for a pyrimidine or a pyrimidine for a purine (C/T "! A/G). An example of a transversion is adenine (A) being converted into a cytosine (C). A point mutation can be reversed by another point mutation, in which the nucleotide is changed back to its original state (true reversion) or by second-site reversion (a complementary mutation elsewhere that results in regained gene functionality). Point mutations that occur within the protein coding region of a gene may be classified into three kinds, depending upon what the erroneous codon codes for:
    - Silent mutations: which code for the same amino acid.
    - Missense mutations: which code for a different amino acid.

- Nonsense mutations: which code for a stop and can truncate the protein.

o Insertions add one or more extra nucleotides into the DNA. They are usually caused by transposable elements, or errors during replication of repeating elements (e.g. AT repeats). Insertions in the coding region of a gene may alter splicing of the mRNA (splice site mutation), or cause a shift in the reading frame (frameshift), both of which can significantly alter the gene product. Insertions can be reverted by excision of the transposable element.

o Deletions remove one or more nucleotides from the DNA. Like insertions, these mutations can alter the reading frame of the gene. They are generally irreversible: though exactly the same sequence might theoretically be restored by an insertion, transposable elements able to revert a very short deletion (say 1–2 bases) in *any* location are either highly unlikely to exist or do not exist at all. Note that a deletion is not the exact opposite of an insertion: the former is quite random while the latter consists of a specific sequence inserting at locations that are not entirely random or even quite narrowly defined.

- Large-scale mutations in chromosomal structure, including:

o Amplifications (or gene duplications) leading to multiple copies of all chromosomal regions, increasing the dosage of the genes located within them.

o Deletions of large chromosomal regions, leading to loss of the genes within those regions.

o Mutations whose effect is to juxtapose previously separate pieces of DNA, potentially bringing together separate genes to form functionally distinct fusion genes (e.g. bcr-abl). These include:

- Chromosomal translocations: interchange of genetic parts from nonhomologous chromosomes.

- Interstitial deletions: an intra-chromosomal deletion that removes a segment of DNA from a single chromosome, thereby apposing previously distant genes. For example, cells isolated from a human astrocytoma, a type of brain tumor, were found to have a chromosomal deletion removing sequences between the "fused in glioblastoma" (fig) gene and the receptor tyrosine kinase "ros", producing a fusion protein (FIG-ROS). The abnormal FIG-ROS fusion protein has constitutively active kinase activity that causes oncogenic transformation (a transformation from normal cells to cancer cells).

  - Chromosomal inversions: reversing the orientation of a chromosomal segment.
- o Loss of heterozygosity: loss of one allele, either by a deletion or recombination event, in an organism that previously had two different alleles.

By Effect on Function

- Loss-of-function mutations are the result of gene product having less or no function. When the allele has a complete loss of function (null allele) it is often called an amorphic mutation. Phenotypes associated with such mutations are most often recessive. Exceptions are when the organism is haploid, or when the reduced dosage of a normal gene product is not enough for a normal phenotype (this is called haploinsufficiency).
- Gain-of-function mutations change the gene product such that it gains a new and abnormal function. These mutations usually have dominant phenotypes. Often called a neomorphic mutation.
- Dominant negative mutations (also called antimorphic mutations) have an altered gene product that acts antagonistically to the wild-type allele. These mutations usually result in an altered molecular function (often inactive) and are characterised by a dominant or semi-dominant phenotype. In humans, Marfan syndrome is an example of a dominant negative mutation occurring in an autosomal dominant disease. In this condition, the defective glycoprotein product of the fibrillin gene (FBN1) antagonizes the product of the normal allele.
- Lethal mutations are mutations that lead to the death of the organisms which carry the mutations.
- A back mutation or reversion is a point mutation that restores the original sequence and hence the original phenotype.

By Effect on Fitness

In applied genetics it is usual to speak of mutations as either harmful or beneficial.

- A harmful mutation is a mutation that decreases the fitness of the organism.
- A beneficial mutation is a mutation that increases fitness of the organism, or which promotes traits that are desirable.

In theoretical population genetics, it is more usual to speak of such mutations as deleterious or advantageous. In the neutral theory of molecular evolution, genetic drift is the basis for most variation at the molecular level.

- A neutral mutation has no harmful or beneficial effect on the organism. Such mutations occur at a steady rate, forming the basis for the molecular clock.
- A deleterious mutation has a negative effect on the phenotype, and thus decreases the fitness of the organism.
- An advantageous mutation has a positive effect on the phenotype, and thus increases the fitness of the organism.
- A nearly neutral mutation is a mutation that may be slightly deleterious or advantageous, although most nearly neutral mutations are slightly deleterious.

In reality, viewing the fitness effects of mutations in these discrete categories is an oversimplification. Attempts have been made to infer the distribution of fitness effects using mutagenesis experiments or theoretical models applied to molecular sequence data. However, the current distribution is still uncertain, and some aspects of the distribution likely vary between species.

By Inheritance

- inheritable generic in pro-generic tissue or cells on path to be changed to gametes.
- non inheritable somatic (e.g., carcinogenic mutation)
- non inheritable post mortem aDNA mutation in decaying remains.

By pattern of inheritance The human genome contains two copies of each gene – a paternal and a maternal allele.

- A heterozygous mutation is a mutation of only one allele.
- A homozygous mutation is an identical mutation of both the paternal and maternal alleles.
- Compound heterozygous mutations or a genetic compound comprises two different mutations in the paternal and maternal alleles.
- A wildtype or homozygous non-mutated organism is one in which neither allele is mutated. (Just not a mutation)

By Impact on Protein Sequence

- A frameshift mutation is a mutation caused by insertion or deletion of a number of nucleotides that is not evenly divisible by three from a DNA sequence. Due to the triplet nature of gene expression by codons, the insertion or deletion can disrupt the reading frame, or the grouping of the codons, resulting in a completely different translation from the original. The earlier in the sequence the deletion or insertion occurs, the more altered the protein produced is.

- A nonsense mutation is a point mutation in a sequence of DNA that results in a premature stop codon, or a *nonsense codon* in the transcribed mRNA, and possibly a truncated, and often nonfunctional protein product.
- Missense mutations or *nonsynonymous mutations* are types of point mutations where a single nucleotide is changed to cause substitution of a different amino acid. This in turn can render the resulting protein nonfunctional. Such mutations are responsible for diseases such as Epidermolysis bullosa, sickle-cell disease, and SOD1 mediated ALS (Boillée 2006, p. 39).
- A neutral mutation is a mutation that occurs in an amino acid codon which results in the use of a different, but chemically similar, amino acid. The similarity between the two is enough that little or no change is often rendered in the protein. For example, a change from AAA to AGA will encode arginine, a chemically similar molecule to the intended lysine. Neutral mutations occur because of the degenerate nature of the genetic code.
- Silent mutations are mutations that do not result in a change to the amino acid sequence of a protein. They may occur in a region that does not code for a protein, or they may occur within a codon in a manner that does not alter the final amino acid sequence. The phrase *silent mutation* is often used interchangeably with the phrase *synonymous mutation*; however, synonymous mutations are a subcategory of the former, occurring only within exons. The name silent could be a misnomer. For example, a silent mutation in the exon/intron border may lead to alternative splicing by changing the splice site, thereby leading to a changed protein.

Special Classes

- Conditional mutation is a mutation that has wild-type (or less severe) phenotype under certain "permissive" environmental conditions and a mutant phenotype under certain "restrictive" conditions. For example, a temperature-sensitive mutation can cause cell death at high temperature (restrictive condition), but might have no deleterious consequences at a lower temperature (permissive condition).

Nomenclature

A committee of the Human Genome Variation Society (HGVS) has developed the standard human sequence variant nomenclature, which should be used by researchers and DNA diagnostic centers to generate unambiguous mutation descriptions. In principle, this nomenclature can also be used to describe mutations in other organisms. The nomenclature specifies the type of mutation and base or amino acid changes.

- Nucleotide substitution (e.g. 76A>T) - The number is the position of the nucleotide from the 5' end, the first letter represents the wild type nucleotide, and the second letter represents the nucleotide which replaced the wild type. In the given example, the adenine at the 76th position was replaced by a thymine.
  - If it becomes necessary to differentiate between mutations in genomic DNA, mitochondrial DNA, and RNA, a simple convention is used. For example, if the 100th base of a nucleotide sequence mutated from G to C, then it would be written as g.100G>C if the mutation occurred in genomic DNA, m.100G>C if the mutation occurred in mitochondrial DNA, or r.100g>c if the mutation occurred in RNA. Note that for mutations in RNA, the nucleotide code is written in lower case.
- Amino acid substitution (e.g. D111E) – The first letter is the one letter code of the wild type amino acid, the number is the position of the amino acid from the N terminus, and the second letter is the one letter code of the amino acid present in the mutation. Nonsense mutations are represented with an X for the second amino acid (e.g. D111X).
- Amino acid deletion (e.g. ÄF508) – The Greek letter Ä (delta) indicates a deletion. The letter refers to the amino acid present in the wild type and the number is the position from the N terminus of the amino acid were it to be present as in the wild type.

The complete set of rules and more examples of mutation descriptions can be found at the HGVS sequence variant nomenclature website. Since the nomenclature has to cover all sequence variants, descriptions can become very complex. To prevent mistakes and facilitate correct use of this nomenclature, the journal Human Mutation recommends the use of Mutalyzer, which can apply the HGVS human nomenclature guidelines to check and, if necessary, correct sequence variant descriptions.

Harmful Mutations

Changes in DNA caused by mutation can cause errors in protein sequence, creating partially or completely non-functional proteins. To function correctly, each cell depends on thousands of proteins to function in the right places at the right times. When a mutation alters a protein that plays a critical role in the body, a medical condition can result. A condition caused by mutations in one or more genes is called a genetic disorder. Some mutations alter a gene's DNA base sequence but do not change the function of the protein made by the gene. Studies of the fly *Drosophila melanogaster* suggest that if a mutation does change a protein, this will probably be harmful, with about 70 percent of these mutations having damaging effects, and the remainder being either neutral or weakly beneficial. However, studies in yeast have shown that only 7% of mutations that are not in genes are harmful. If a mutation is present in a germ cell, it can give rise to offspring that carries the mutation in all of its cells. This is the case in hereditary diseases. On the other hand, a mutation may occur in a somatic cell of an organism. Such mutations will be present in all descendants of this cell within the same organism, and certain mtations can cause the cell to become malignant, and thus cause cncer. Often, gene mutations that could cause a genetic disorder are repaired by the DNA repair system of the cell. Each cell has a number of pathways through which enzymes recognize and repair mistakes in DNA. Because DNA can be damaged or mutated in many ways, the process of DNA repair is an important way in which the body protects itself from disease.

Although most mutations that change protein sequences are neutral or harmful, some mutations have a positive effect on an organism. In this case, the mutation may enable the mutant organism to withstand particular environmental stresses better than wild-type organisms, or reproduce more quickly. In these cases a mutation will tend to become more common in a population through natural selection. For example, a specific 32 base pair deletion in human CCR5 (CCR5-Ä32) confers HIV resistance to homozygotes and delays AIDS onset in heterozygotes. The CCR5 mutation is more common in those of European descent. One possible explanation of the etiology of the relatively high frequency of CCR5-Ä32 in the European population is that it conferred resistance to the bubonic plague in mid-14th century Europe. People with this mutation were more likely to survive infection; thus its frequency in the population increased. This theory could explain why this mutation is not found in southern Africa, where the bubonic plague never reached. A newer theory suggests that the selective pressure on the CCR5 Delta 32 mutation was caused by smallpox instead of the bubonic plague.

Another example is Sickle cell disease, a blood disorder in which the body produces an abnormal type of the oxygen-carrying substance hemoglobin in the red blood cells. One-third of all indigenous inhabitants of Sub-Saharan Africa carry the gene, because in areas where malaria is common, there is a survival value in carrying only a single sickle-cell gene (sickle cell trait). Those with only one of the two alleles of the sickle-cell disease are more resistant to malaria, since the infestation of the malaria plasmodium is halted by the sickling of the cells which it infests.

Prion Mutation

Prions are proteins and do not contain genetic material. However, prion replication has een shown to be subject to mutation and natural selection just like other forms of replication.

Genetic Variation

Genetic variation, variation in alleles of genes, occurs both within and among populations. Genetic variation is important because it provides the "raw material" for natural selection. Genetic variation is brought about by mutation, a change in a chemical structure of a gene. Polyploidy is an example of chromosomal mutation. Polyploidy is a condition wherein organisms have three or more sets of genetic variation (3n or more).

Among Individuals Within a Population

Genetic variation among individuals within a population can be identified at a variety of levels. It is possible to identify genetic variation from observations of phenotypic variation in either quantitative traits (traits that vary continuously and are coded for by many genes, e.g., leg length in dogs) or discrete traits (traits that fall into discrete categories and are coded for by one or a few genes (e.g., white, pink, red petal color in certain flowers). Genetic variation can also be identified by examining variation at the level of enzymes using the process of protein electrophoresis. Polymorphic genes have more than one allele at each locus. Half of the genes that code for enzymes in insects and plants may be polymorphic, whereas polymorphisms are less common in vertebrates. Ultimately, genetic variation is caused by variation in the order of bases in the nucleotides in genes. New technology now allows scientists to directly sequence DNA which has identified even more genetic variation than was previously detected by protein electrophoresis. Examination of DNA has sown genetic variation in both coding regions and in the non-coding intron region of genes. Genetic variation will result in phenotypic variation if variation in the order of nucleotides in the DNA sequence results in a difference in the order of amino acids in proteins coded by that DNA sequence, and if the resultant differences in amino acid sequence influence the shape, and thus the function of the enzyme.

Between Populations

Geographic variation in genes often occurs among populations living in different locations. Geographic variation may be due to differences in selective pressures or to genetic drift.

Measurement

Genetic variation within a population is commonly measured as the percentage of gene loci that are polymorphic or the percentage of gene loci in individuals that are heterozygous.

Sources

Mutations are the ultimate source of genetic variation because they alter the order of bases in the nucleotides of DNA. Mutations are likely to be rare and most mutations are neutral or deleterious, but in some instances the new alleles can be favoured by natural selection.

Genetic variation can also be produced by the recombination of chromosomes that occurs during sexual reproduction, called independent assortment.

Crossing over and random segregation during meiosis can result in the production of new alleles or new combinations of alleles. Furthermore, random ferilisation also contributes to variation.

Variation and recombination can be fascilitated by transposable and transposed genetic elements, commonly known as endogenous retroviruses, LINEs, SINEs, etc.

Maintenance in Populations

A variety of factors maintain genetic variation in populations. Potentially harmful recessive alleles can be hidden from selection in the heterozygous individuals in populations of diploid organisms (recessive alleles are only expressed in the less common homozygous individuals). Natural selection can also maintain genetic variation in balanced polymorphisms. Balanced polymorphisms may occur when heterozygotes are favoured or when selection is frequency dependent.

8

Genetic Genealogy

Genetic genealogy is the application of genetics to traditional genealogy. Genetic genealogy involves the use of genealogical DNA testing to determine the level of genetic relationship between individuals. The investigation of surnames in genetics can be said to go back to George Darwin, a son of Charles Darwin. In 1875, George Darwin used surnames to estimate the frequency of first-cousin marriages and calculated the expected incidence of marriage between people of the same surname (isonymy). He arrived at a figure between 2.25% and 4.5% for cousin-marriage in the population of Great Britain, with the upper classes being on the high end and the general rural population on the low end. (His parents, Charles Darwin and Emma Wedgwood, were first cousins.) This simple study was innovative for its era. The next stimulus toward using genetics to study family history had to wait until the 1990s, when certain locations on the Y chromosome were identified as being useful for tracing male-to-male inheritance. Dr. Karl Skorecki, a Canadian nephrologist of Ashkenazi parentage, noticed that a Sephardic fellow-congregant who was a Kohen like himself had completely different physical features. According to Jewish tradition, all Kohanim are descended from the priest Aaron, brother of Moses. Skorecki reasoned that if Kohanim were indeed the descendants of only one man, they should have a common set of genetic markers and should perhaps preserve some family resemblance to each other.

To test that hypothesis, he contacted Professor Michael Hammer of the University of Arizona, a researcher in molecular genetics and pioneer in Y chromosome research. Their report in the *Nature* in 1997 sent shock waves through the worlds of science and religion. A particular marker was indeed more likely to be present in Jewish men from the priestly tradition than in the general Jewish population. It was apparently true that a common descent had been strictly preserved for thousands of years. Moreover, the data showed that there were very few "non-paternity events". The first to test the new methodology in general surname research was Bryan Sykes, a molecular biologist at Oxford University. His study of the Sykes surname obtained valid results by looking at only four markers on the male chromosome. It pointed the way to genetics becoming a valuable assistant in the service of genealogy and history.

In April 2000, Family Tree DNA began offering the first genetic genealogy tests to the public. This offering marked the first time that a personal theory on the Y chromosome could be tested outside of an academic study. Additionally, Sykes' concept of a surname study, which by this time had been adopted by several other academic researchers outside of Oxford, was expanded into online Surname Projects (an early form of social network) and the effort helped spread knowledge gained through testing to interested genealogists worldwide.

In 2001, Sykes went on to write the popular book *The Seven Daughters of Eve*, which described the seven major haplogroups of European ancestors. In the wake of the book's success, and with the growing availability and affordability of genealogical DNA testing, genetic genealogy as a field began growing rapidly. By 2003, the field of DNA testing of surnames was declared officially to have "arrived" in an article by Jobling and Tyler-Smith in *Nature Reviews Genetics*. The number of firms offering tests, and the number of consumers ordering them, had risen dramatically. Another milestone in the acceptance of genetic genealogy is the Genographic Project. The Genographic Project is a five-year research study launched in 2005 by the National Geographic Society and IBM, in partnership with the University of Arizona and Family Tree DNA. Although its goals are primarily anthropological, not genealogical, the project's sale by April 2010 of more than 350,000 of its public participation testing kits, which test the general public for either twelve STR markers on the Y chromosome or mutations on the HVR1 region of the mtDNA, has helped increase the visibility of genetic genealogy.

More state-of-the-art commercial laboratories now recommend testing at least 25 markers, since the more markers tested, the more discriminating and powerful the results will be. A 12-marker STR test is usually not discriminating enough to provide conclusive results for a common surname. Genetic laboratories such as Genebase and Family Tree DNA give the option of testing 67 Y-DNA Markers. Annual sales of genetic genealogical tests for all companies, including the laboratories that support them, are estimated to be in the area of $60 million (2006).

Interpretation

Since the year 2000, dozens of relevant academic papers have been published, and thousands of private test results organised by surname study groups have been made available on the internet. The comparison of results may be complicated by the fact that some laboratories use different testing methods. Apparently differing results from two sources may in fact be identical, and vice-versa.

Uses

The two most common types of genetic genealogy tests are Y-DNA (paternal line) and mtDNA (maternal line) genealogical DNA tests. Note that Y chromosome and Y-DNA are used interchangeably in this article. These tests involve the comparison of certain sequences of the DNA of pairs of individuals in order to estimate the probability that they share a common ancestor in a genealogical time frame and, through the use of a Bayesian model published by Bruce Walsh, to estimate the number of generations separating the two individuals from their most recent common ancestor or "mrca". Y-DNA testing involves short tandem repeat (STR) and, sometimes, single nucleotide polymorphism (SNP) testing of the Y-chromosome. The Y-chromosome is present only in males and reveals information on the strict paternal line. These tests can provide insight into the recent (via STRs) and ancient (via SNPs) genetic ancestry. A Y-chromosome STR test will reveal a haplotype, which should be similar among all male descendants of a male ancestor. SNP tests are used to assign people to a paternal haplogroup, which defines a much larger genetic population. mtDNA testing involves sequencing or testing the HVR-1 region, HVR-2 region or both. An mtDNA test may also include the additional SNPs needed to assign people to a maternal haplogroup—or even include the complete mtDNA.

Either Y-DNA or mtDNA test results can be compared to the results of others via private or public DNA databases.

Biogeographical and Ethnic Origins

Additional DNA tests exist for determining biogeographical and ethnic origin, but these tests have less relevance for traditional genealogy.

Genetic genealogy has revealed astonishing links between peoples. For instance, it has shown that the ancient Phoenician people were ancestors of much of the present-day population of the island of Malta. Preliminary results from a study by Pierre Zalloua of the American University of Beirut and Spencer Wells, supported by a grant from National Geographic's Committee for Research and Exploration, were published in the October 2004 issue of *National Geographic*. One of the conclusions is that "more than half of the

Y chromosome lineages that we see in today's Maltese population could have come in with the Phoenicians."

Human Migration

Genealogical DNA testing methods are also being used on a longer time scale to trace human migratory patterns. For example, they have been used to determine when the first humans came to North America and what path they followed. For several years, a number of researchers and laboratories from around the world have been sampling indigenous populations from around the globe in an effort to map historical human migration patterns. Recently, several projects have been created that are aimed at bringing this science to the public. One example, mentioned in History above, is the National Geographic Society's Genographic Project, which aims to map historical human migration patterns by collecting and analyzing DNA samples from over 100,000 people across five continents. Another example is the DNA Clans Genetic Ancestry Analysis, which measures a person's precise genetic connections to indigenous ethnic groups from around the world.

Typical Customers and Interest Groups

Male DNA testing customers most often start with a Y chromosome test to determine their father's paternal ancestry. Females generally begin with a mitochondrial test to trace their ancient maternal lineage, which males often have tested for the same purpose.

A common consumer goal in purchasing DNA testing services is to acquire quantified, scientific linkage to a specific ancestral group. A compelling example of this motive is found in the expressed desires of some consumers to be proven to have Viking paternal ancestry. In keeping with this marketplace demand, one British DNA testing service, Oxford Ancestors, offers a Y chromosome test purporting to assess whether given males are of "Viking stock." Those whose DNA falls into the designated haplogroup are issued Viking Descendant certificates by the testing service. The same DNA testing company participated in producing a televised documentary, "The Blood of the Vikings," in conjunction with the BBC, which showed how DNA testing could reveal Viking ancestry.

The RootsWeb Genealogy-DNA Internet discussion group has a membership of 750 subscribers from around the world. Some subscribers have had various DNA tests performed and are seeking advice and guidance in interpreting their results. The list also includes administrators of DNA projects that examine surnames, geographic regions, or ethnic groups. The sophistication of subscribers ranges from expert to novice. In some cases, subscribers have been credited with making useful and novel contributions to knowledge in the field of genetic genealogy.

Paternal and Maternal DNA Lineages

Mitochondria are small organelles that lie in the cytoplasm of eukaryotic cells, such as those of humans. Their primary purpose is to provide energy to the cell. Mitochondria are thought to be the vestigial remains of symbiotic bacteria that were once free living. One indication that mitochondria were once free living is that they contain a relatively small circular segment of DNA, called mitochondrial DNA (mtDNA). The overwhelming majority of a human's DNA is contained in chromosomes in the nucleus of the cell, but mtDNA is an exception. Individuals inherit their cytoplasm and the organelles it contains exclusively from their mothers, as these are derived from the ovum (egg cell) only, not from the sperm.

When a mutation arises in mtDNA molecule, the mutation is therefore passed in a direct female line of descent. These rare mutations are derived from copying mistakes—when the DNA is copied it is possible that a single mistake occurs in the DNA sequence, an outcome which is called a single nucleotide polymorphism (SNP).

Human Y chromosomes are male-specific sex chromosomes; nearly all humans that possess a Y chromosome will be morphologically male. Y chromosomes are therefore passed from father to son; although Y chromosomes are situated in the cell nucleus, they only recombine with the X chromosome at the ends of the Y chromosome; the vast majority of the Y chromosome (95%) does not recombine. When mutations (SNPs, and STR copying mistakes) arise in the Y chromosome, they are passed down directly from father to son in a direct male line of descent. The Y-DNA and mtDNA therefore share a certain feature: they both pass down unchanged except for mutations.

The other chromosomes, autosomes and X chromosomes in women, share their genetic material (called crossing over leading to recombination) during meiosis (a special type of cell division that occurs for the purposes of sexual reproduction). Effectively this means that the genetic material from these chromosomes gets mixed up in every generation, and so any new mutations are passed down randomly from parents to offspring. The special feature that both Y-DNA and mtDNA share, above, preserves a "written" record of their mutations because neither DNA gets mixed up or randomized—mutations remain fixed in place on both types of DNA. Furthermore the historical sequence of these mutations can also be inferred. For example, if a set of ten Y chromosomes (derived from ten different men) contains a mutation, A, but only five of these chromosomes contain a second mutation, B, it must be the case that mutation B occurred after mutation A.

Furthermore all ten men who carry the chromosome with mutation A are the direct male line descendants of the same man who was the first to carry this mutation. The first man to carry mutation B was also a direct male line descendant of this man, but is also the direct male line ancestor of all men carrying mutation B. Series of mutations such as this form molecular lineages. Furthermore each SNP mutation may define a set of specific Y chromosomes called a haplogroup.

All men carrying SNP mutation A form a single haplogroup, and all men carrying mutation B are part of this haplogroup, but mutation B (if a SNP) may also define a more recent haplogroup (which is a subgroup or subclade) of its own which men carrying only mutation A do not belong to. Both mtDNA and Y chromosomes or Y-DNA are grouped into lineages and haplogroups; these are often presented as tree-like diagrams.

Benefits

Genetic genealogy gives genealogists a means to check or supplement their genealogy results with information obtained via DNA testing. A positive test match with another individual may:

- provide locations for further genealogical research
- help determine ancestral homeland
- discover living relatives
- validate existing research
- confirm or deny suspected connections between families
- prove or disprove theories regarding ancestry.

Drawbacks

People who resist testing may cite one of the following concerns:

- Cost
- Quality of testing
- Concerns over privacy issues
- Loss of ethnic identity.

Finally, Y-DNA and mtDNA tests each only trace a single lineage (one's father's father's father's etc. lineage or one's mother's mother's mother's etc. lineage). At 10 generations back, an individual has up to 1024 unique ancestors (fewer if ancestor cousins interbred) and a Y-DNA or mtDNA test is only studying one of those ancestors, as well as their descendants and siblings (same sexed siblings for Y-DNA or all siblings for mtDNA). However, most genealogists maintain contact with many cousins (1st, 2nd, 3rd, etc., with different surnames) whose Y-DNA and mtDNA are different, and thus can be encouraged to be tested to find additional ancestral DNA lineages.

Expected Growth

Genetic genealogy is a rapidly growing field. As the cost of testing continues to drop, the number of people being tested continues to increase. The probability of finding a genetic match among the DNA databases should continue to improve. Laboratories and testing firms are engaging in active research and development that will allow for higher confidence intervals and better results interpretation, including historical interpretive reports and customized research.

Genetic Distance among Relatives

Where the genogram or family tree of individuals is known, it can be used to determine the genetic identity between individuals. It is often described as percentage of genetic identity, referring to the fraction of genome inherited from common ancestors, and not actual genomic identity, which is always approximately 99.9% identical from one human to another. One method of calculating this genetic similarity is to do an inbreeding calculation by the path or tabular method and then multiply by 2, because any progeny would have a 1 in 2 risk of actually inheriting the identical alleles from both parents. For instance, a brother/sister relation gives 25% risk for two alleles to be identical by descent.

History of Molecular Biology

The history of molecular biology begins in the 1930s with the convergence of various, previously distinct biological disciplines: biochemistry, genetics, microbiology, and virology. With the hope of understanding life at its most fundamental level, numerous physicists and chemists also took an interest in what would become molecular biology.

In its modern sense, molecular biology attempts to explain the phenomena of life starting from the macromolecular properties that generate them. Two categories of macromolecules in particular are the focus of the molecular biologist: 1) nucleic acids, among which the most famous is deoxyribonucleic acid (or DNA), the constituent of genes, and 2) proteins, which are the active agents of living organisms. One definition of the scope of molecular biology therefore is to characterize the structure, function and relationships between these two types of macromolecules. This relatively limited definition will suffice to allow us to establish a date for the so-called "molecular revolution", or at least to establish a chronology of its most fundamental developments.

General Overview

In its earliest manifestations, molecular biology—the name was coined by Warren Weaver of the Rockefeller Foundation in 1938—was an ideal of physical and chemical explanations of life, rather than a coherent discipline. Following the advent of the Mendelian-chromosome theory of heredity in the 1910s and the maturation of atomic theory and quantum mechanics in the 1920s, such explanations seemed within reach. Weaver and others encouraged (and funded) research at the intersection of biology, chemistry and physics, while prominent physicists such as Niels Bohr and Erwin Schrodinger turned their attention to biological speculation. However, in the 1930s and 1940s it was by no means clear which—if any—cross-disciplinary research would bear fruit; work in colloid chemistry, biophysics and radiation biology, crystallography, and other emerging fields all seemed promising.

In 1940, George Beadle and Edward Tatum demonstrated the existence of a precise relationship between genes and proteins. In the course of their experiments connecting genetics with biochemistry, they switched from the genetics mainstay *Drosophila* to a more appropriate model organism, the fungus *Neurospora*; the construction and exploitation of new model organisms would become a recurring theme in the development of molecular biology. In 1944, Oswald Avery, working at the Rockefeller Institute of New York, demonstrated that genes are made up of DNA. In 1952, Alfred Hershey and Martha Chase confirmed that the genetic material of the bacteriophage, the virus which infects bacteria, is made up of DNA. In 1953, James Watson and Francis Crick discovered the double helical structure of the DNA molecule. In 1961, Francois Jacob and Jacques Monod hypothesized the existence of an intermediary between DNA and its protein products, which they called messenger RNA. Between 1961 and 1965, the relationship between the information contained in DNA and the structure of proteins was determined: there is a code, the genetic code, which creates a correspondence between the succession of nucleotides in the DNA sequence and a series of amino acids in proteins. At the beginning of the 1960s, Monod and Jacob also demonstrated how certain specific proteins, called regulative proteins, latch onto DNA at the edges of the genes and control the transcription of these genes into messenger RNA; they direct the "expression" of the genes.

The chief discoveries of molecular biology took place in a period of only about twenty-five years. Another fifteen years were required before new and more sophisticated technologies, united today under the name of genetic engineering, would permit the isolation and characterization of genes, in particular those of highly complex organisms.

The Exploration of the Molecular Dominion

If we evaluate the molecular revolution within the context of biological history, it is easy to note that it is the culmination of a long process which began with the first observations through a microscope. The aim of these early researchers was to understand the functioning of living organisms by describing their organization at the microscopic level. From the end of the 18th century, the characterization of the chemical molecules which make up living beings gained increasingly greater attention, along with the birth of physiological chemistry in the 19th century, developed by the German chemist Justus von Liebig and following the birth of biochemistry at the beginning of the 20th, thanks to another German chemist Eduard Buchner. Between the molecules studied by chemists and the tiny structures visible under the optical microscope, such as the cellular nucleus or the chromosomes, there was an obscure zone, "the world of the ignored dimensions," as it was called by the chemical-physicist Wolfgang Ostwald. This world is populated by colloids, chemical compounds whose structure and properties were not well defined.

The successes of molecular biology derived from the exploration of that unknown world by means of the new technologies developed by chemists and physicists: X-ray diffraction, electron microscopy, ultracentrifugization, and electrophoresis. These studies revealed the structure and function of the macromolecules.

A milestone in that process was the work of Dr. Linus Pauling in 1949, which for the first time linked the specific genetic mutation in patients with sickle cell disease to a demonstrated change in an individual protein, the hemoglobin in the erythrocytes of heterozygous or homozygous individuals.

The Encounter between Biochemistry and Genetics

The development of molecular biology is also the encounter of two disciplines which made considerable progress in the course of the first thirty years of the twentieth century: biochemistry and genetics. The first studies the structure and function of the molecules which make up living things. Between 1900 and 1940, the central processes of metabolism were described: the process of digestion and the absorption of the nutritive elements derived from alimentation, such as the sugars. Every one of these processes is catalysed by a particular enzyme. Enzymes are proteins, like the antibodies present in blood or the proteins responsible for muscular contraction. As a consequence, the study of proteins, of their structure and synthesis, became one of the principal objectives of biochemists.

The second discipline of biology which developed at the beginning of the 20th century is genetics. After the rediscovery of the laws of Mendel through the studies of Hugo de Vries, Carl Correns and Erich von Tschermak in 1900, this science began to take shape thanks to the adoption by Thomas Hunt Morgan, in 1910, of a model organism for genetic studies, the famous fruit fly (Drosophila melanogaster). Shortly after, Morgan showed that the genes are localized on chromosomes. Following this discovery, he continued working with Drosophila and, along with numerous other research groups, confirmed the importance of the gene in the life and development of organisms. Nevertheless, the chemical nature of genes and their mechanisms of action remained a mystery. Molecular biologists committed themselves to the determination of the structure, and the description of the complex relations between, genes and proteins. The development of molecular biology was not just the fruit of some sort of intrinsic "necessity" in the history of ideas, but was a characteristically historical phenomenon, with all of its unknowns, imponderables and contingencies: the remarkable developments in physics at the beginning of the 20th century highlighted the relative lateness in development in biology, which became the "new frontier" in the search for knowledge about the empirical world. Moreover, the developments of the theory of information and cybernetics in the 1940s, in response to military exigencies, brought to the new biology a significant number of fertile ideas and, especially, metaphors.

The choice of bacteria and of its virus, the bacteriophage, as models for the study of the fundamental mechanisms of life was almost natural - they are the smallest living organisms known to exist - and at the same time the fruit of individual choices. This model owes its success, above all, to the fame and the sense of organization of Max Delbruck, a German physicist, who was able to create a dynamic research group, based in the United States, whose exclusive scope was the study of the bacteriophage: the *School of the Phage*.

The geographic panorama of the developments of the new biology was conditioned above all by preceding work. The US, where genetics had developed the most rapidly, and the UK, where there was a coexistence of both genetics and biochemical research of highly advanced levels, were in the avant-garde. Germany, the cradle of the revolutions in physics, with the best minds and the most advanced laboratories of genetics in the world, should have had a primary role in the development of molecular biology. But history decided differently: the arrival of the Nazis in 1933 - and, to a less extreme degree, the rigidification of totalitarian measures in fascist Italy - caused the emigration of a large number of Jewish and non-Jewish scientists. The majority of them fled to the US or the UK, providing an extra impulse to the scientific dynamism of those nations. These movements ultimately made molecular biology a truly international science from the very beginnings.

DNA Biochemistry

First Isolation of DNA

Working in the 19th century, biochemists initially isolated DNA and RNA (mixed together) from cell nuclei. They were relatively quick to appreciate the polymeric nature of their "nucleic acid" isolates, but realized only later that nucleotides were of two types—one containing ribose and the other deoxyribose. It was this subsequent discovery that led to the identification and naming of DNA as a substance distinct from RNA. Friedrich Miescher (1844–1895) discovered a substance he called "nuclein" in 1869. Somewhat later, he isolated a pure sample of the material now known as DNA from the sperm of salmon, and in 1889 his pupil, Richard Altmann, named it "nucleic acid". This substance was found to exist only in the chromosomes.

In 1919 Phoebus Levene at the Rockefeller Institute identified the components (the four bases, the sugar and the phosphate chain) and he showed that the components of DNA were linked in the order phosphate-sugar-base. He called each of these units a nucleotide and suggested the DNA molecule consisted of a string of nucleotide units linked together through the phosphate groups, which are the 'backbone' of the molecule. However Levene thought the chain was short and that the bases repeated in the same fixed order. Torbjorn Caspersson and Einar Hammersten showed that DNA was a polymer.

Chromosomes and Inherited Traits

Max Delbruck, Nikolai V. Timofeeff-Ressovsky, and Karl G. Zimmer published results in 1935 suggesting that chromosomes are very large molecules the structure of which can be changed by treatment with X-rays, and that by so changing their structure it was possible to change the heritable characteristics governed by those chromosomes. In 1937 William Astbury produced the first X-ray diffraction patterns from DNA. He was not able to propose the correct structure but the patterns showed that DNA had a regular structure and therefore it might be possible to deduce what this structure was.

In 1943, Oswald Theodore Avery and a team of scientists discovered that traits proper to the "smooth" form of the *Pneumococcus* could be transferred to the "rough" form of the same bacteria merely by making the killed "smooth" (S) form available to the live "rough" (R) form. Quite unexpectedly, the living R *Pneumococcus* bacteria were transformed into a new strain of the S form, and the transferred S characteristics turned out to be heritable. Avery called the medium of transfer of traits the transforming principle; he identified DNA as the transforming principle, and not protein as previously thought. He essentially redid Frederick Griffith's experiment. In 1953, Alfred Hershey and Martha Chase did an experiment (Hershey-Chase experiment) that showed, in T2 phage, that DNA is the genetic material (Hershey shared the Nobel prize with Luria).

Structure of DNA Discovery

In the 1950s, three groups made it their goal to determine the structure of DNA. The first group to start was at King's College London and was led by Maurice Wilkins and was later joined by Rosalind Franklin. Another group consisting of Francis Crick and James D. Watson was at Cambridge. A third group was at Caltech and was led by Linus Pauling. Crick and Watson built physical models using metal rods and balls, in which they incorporated the known chemical structures of the nucleotides, as well as the known position of the linkages joining one nucleotide to the next along the polymer. At King's College Maurice Wilkins and Rosalind Franklin examined X-ray diffraction patterns of DNA fibres. Of the three groups, only the London group was able to produce good quality diffraction patterns and thus produce sufficient quantitative data about the structure.

Helix Structure

In 1948 Pauling discovered that many proteins included helical shapes. Pauling had deduced this structure from X-ray patterns and from attempts to physically model the structures. (Pauling was also later to suggest an incorrect three chain helical DNA structure based on Astbury's data.) Even in the initial diffraction data from DNA by Maurice Wilkins, it was evident that the structure involved helices. But this insight was only a beginning. There remained the questions of how many strands came together, whether this number was the same for every helix, whether the bases pointed toward the helical axis or away, and ultimately what were the explicit angles and coordinates of all the bonds and atoms. Such questions motivated the modelling efforts of Watson and Crick.

Complementary Nucleotides

In their modelling, Watson and Crick restricted themselves to what they saw as chemically and biologically reasonable. Still, the breadth of possibilities was very wide. A breakthrough occurred in 1952, when Erwin Chargaff visited Cambridge and inspired Crick with a description of experiments Chargaff had published in 1947. Chargaff had observed that the proportions of the four nucleotides vary between one DNA sample and the next, but that for particular pairs of nucleotides — adenine and thymine, guanine and cytosine — the two nucleotides are always present in equal proportions. Using X-ray diffraction, as well as other data from Rosalind Franklin and her information that the bases were paired, James D. Watson and Francis Crick arrived at the first accurate model of DNA's molecular structure in 1953, which was accepted through inspection by Rosalind Franklin.

The discovery was made on February 28, 1953; the first Watson/Crick paper appeared in Nature on April 25, 1953. Sir Lawrence Bragg, the director of the Cavendish Laboratory, where Watson and Crick worked, gave a talk at Guys Hospital Medical School in London on Thursday, May 14, 1953 which resulted in an article by Ritchie Calder in The News Chronicle of London, on Friday, May 15, 1953, entitled "Why You Are You. Nearer Secret of Life." The news reached readers of The New York Times the next day; Victor K. McElheny, in researching his biography, "Watson and DNA: Making a Scientific Revolution", found a clipping of a six-paragraph New York Times article written from London and dated May 16, 1953 with the headline "Form of `Life Unit' in Cell Is Scanned." The article ran in an early edition and was then pulled to make space for news deemed more important.(The New York Times subsequently ran a longer article on June 12, 1953). The Cambridge University undergraduate newspaper also ran its own short article on the discovery on Saturday, May 30, 1953. Bragg's original announcement at a Solvay conference on proteins in Belgium on 8 April 1953 went unreported by the press! In 1962 Watson, Crick, and Maurice Wilkins jointly received the Nobel Prize for Physiology or Medicine for their determination of the structure of DNA.

"Central Dogma"

Watson and Crick's model attracted great interest immediately upon its presentation. Arriving at their conclusion on February 21, 1953, Watson and Crick made their first announcement on February 28. In an influential presentation in 1957, Crick laid out the "Central Dogma", which foretold the relationship between DNA, RNA, and proteins, and articulated the "sequence hypothesis." A critical confirmation of the replication mechanism that was implied by the double-helical structure followed in 1958 in the form of the Meselson-Stahl experiment. Work by Crick and coworkers showed that the genetic code was based on non-overlapping triplets of bases, called codons, and Har Gobind Khorana and others deciphered the genetic code not long afterward (1966). These findings represent the birth of molecular biology.

RNA Tertiary Structure History

The Helical Structure of RNA

The earliest work in RNA structural biology coincided, more or less, with the work being done on DNA in the early 1950s. In their seminal 1953 paper, Watson and Crick suggested that van der Waals crowding by the 2'OH group of ribose would preclude RNA from adopting a double helical structure identical to the model they proposed - what we now know as B-form DNA. This provoked questions about the three dimensional structure of RNA: could this molecule form some type of helical structure, and if so, how? As with DNA, early structural work on RNA centered around isolation of native RNA polymers for fibre diffraction analysis. In part because of heterogeneity of the samples tested, early fibre diffraction patterns were usually ambiguous and not readily interpretable. In 1955, Grunberg-Manago *et al.* published a paper describing the enzyme polynucleotide phosphorylase, which cleaved a phosphate group from nucleotide diphosphates to catalyse their polymerization. This discovery allowed researchers to synthesize homogenous nucleotide polymers, which they then combined to produce double stranded molecules. These samples yielded the most readily interpretable fibre diffraction patterns yet obtained, suggesting an ordered, helical structure for cognate, double stranded RNA that differed from that observed in DNA. These results paved the way for a series of investigations into the various properties and propensities of RNA. Through the late 1950s and early 1960s, numerous papers were publihed on various topics in RNA structure, including RNA-DNA hybridization, triple stranded RNA, and even small-scale crystallography of RNA di-nucleotides - G-C, and A-U - in primitive helix-like arrangements. For a more in-depth review of the early work in RNA structural biology.

The Beginning: Crystal Structure of tRNA

In the mid-1960s, the role of tRNA in protein synthesis was being intensively studied. At this point, ribosomes had been implicated in protein synthesis, and it had been shown that an mRNA strand was necessary for the formation of these structures. In a 1964 publication, Warner and Rich showed that ribosomes active in protein synthesis contained tRNA molecules bound at the A and P sites, and discussed the notion that these molecules aided in the peptidyl transferase reaction. However, despite considerable biochemical characterization, the structural basis of tRNA function remained a mystery. In 1965, Holley *et al.* purified and sequenced the first tRNA molecule, initially proposing that it adopted a cloverleaf structure, based largely on the ability of certain regions of the molecule to form stem loop structures. The isolation of tRNA proved to be the first major windfall in RNA structural biology. Following Holley's publication, numerous investigators began work on isolation tRNA for crystallographic study, developing improved methods for isolating the molecule as they worked. By 1968 several groups had produced tRNA crystals, but these proved to be of limited quality and did not yield data at the resolutions necessary to determine structure. In 1971, Kim *et al.* achieved another breakthrough, producing crystals of yeast tRNA that diffracted to 2-3 Angstrom resolutions by using spermine, a naturally occurring polyamine, which bound to and stabilized the tRNA. Despite having suitable crystals, however, the structure of tRNA was not immediately solved at high resolution; rather it took pioneering work in the use of heavy metal derivatives and a good deal more time to produce a high-quality density map of the entire molecule. In 1973, Kim *et al.* produced a 4 Angstrom map of the tRNA molecule in which they could unambiguously trace the entire backbone. This solution would be followed by many more, as various investigators worked to refine the structure and thereby more thoroughly elucidate the details of base pairing and stacking interactions, and validate the published architecture of the molecule.

The tRNA structure is notable in the field of nucleic acid structure in general, as it represented the first solution of a long-chain nucleic acid structure of any kind - RNA or DNA - preceding Dickerson's solution of a B-form dodecamer by nearly a decade. Also, tRNA demonstrated many of the tertiary interactions observed in RNA architecture which would not be categorized and more thoroughly understood for years to come, providing a foundation for all future RNA structural research.

The Renaissance: The Hammerhead Ribozyme and the Group I Intron: P_{4-6}

For a considerable time following the first tRNA structures, the field of RNA structure did not dramatically advance. The ability to study an RNA strcuture depended upon the potential to isolate the RNA target. This proved limiting to the field for many years, in part owing to the fact that other know targets - i.e. the ribosome - were significantly more difficult to isolate and crystallize. Further, because other interesting RNA targets had simply not been identified, or were not sufficiently understood to be deemed interesting, there was simply a lack of things to study structurally. As such, for some twenty years following the original publication of the tRNA structure, the structures of only a handful of other RNA targets were solved, with almost all of these belonging to the transfer RNA family. This unfortunate lack of scope would eventually be overcome largely because of two major advancements in nucleic acid research: the identification of ribozymes, and the ability to produce them via *in vitro* transcription.

Subsequent to Tom Cech's publication implicating the *Tetrahymena* group I intron as an autocatalytic ribozyme, and Sidney Altman's report of catalysis by ribonuclease P RNA, several other catalytic RNAs were identified in the late 1980s, including the hammerhead ribozyme. In 1994, McKay *et al.* published the structure of a 'hammerhead RNA-DNA ribozyme-inhibitor complex' at 2.6 Angstrom resolution, in which the autocatalytic activity of the ribozyme was disrupted via binding to a DNA substrate. The conformation of the ribozyme published in this paper was eventually shown to be one of several possible states, and although this particular sample was catalytically inactive, subsequent structures have revealed its active-state architecture. This structure was followed by Doudna's publication of the structure of the P4-P6 domains of the *Tetrahymena* group I intron, a fragment of the ribozyme originally made famous by Cech. The second clause in the title of this publication - *Principles of RNA Packing* - concisely evinces the value of these two structures: for the first time, comparisons could be made between well described tRNA structures and those of globular RNAs outside the transfer family. This allowed the framework of categorization to be built for RNA tertiary structure. It was now possible to propose the conservation of motifs, folds, and various local stabilizing interactions. For an early review of these structures and their implications.

In addition to the advances being made in global structure determination via crystallography, the early 1990s also saw the implementation of NMR as a powerful technique in RNA structural biology. Coincident with the large-scale ribozyme structures being solved crystallographically, a number of structures of small RNAs and RNAs complexed with drugs and peptides were solved using NMR. In addition, NMR was now being used to investigate and supplement crystal structures, as exemplified by the determination of an isolated tetraloop-receptor motif structure published in 1997. Investigations such as this enabled a more precise characterization of the base pairing and base stacking interactions which stabilized the global folds of large RNA molecules. The importance of understanding RNA tertiary structural motifs was prophetically well described by Michel and Costa in their publication identifying the tetraloop motif: "..it should not come as a surprise if self-folding RNA molecules were to make intensive use of only a relatively small set of tertiary motifs. Identifying these motifs would greatly aid modelling enterprises, which will remain essential as long as the crystallization of large RNAs remains a difficult task".

The Modern Era: The Age of RNA Structural Biology

The resurgence of RNA structural biology in the mid-1990s has caused a veritable explosion in the field of nucleic acid structural research. Since the publication of the hammerhead and $P_{4\text{-}6}$ structures, numerous major contributions to the field have been made. Some of the most noteworthy examples include the structures of the Group I and Group II introns, and the Ribosome. It should be noted that the first three structures were produced using *in vitro* transcription, and that NMR has played a role in investigating partial components of all four structures - testaments to the indispensability of both techniques for RNA research. Most recently, the 2009 Nobel Prize in Chemistry was awarded to Ada Yonath, Venkatraman Ramakrishnan, and Thomas Steitz for their structural work on the ribosome, demonstrating the prominent role RNA structural biology has taken in modern molecular biology.

Protein Biochemistry

First Isolation and Classification

Proteins were recognized as a distinct class of biological molecules in the eighteenth century by Antoine Fourcroy and others. Members of this class (called the "albuminoids", *Eiweisskorper*, or *matières albuminoides*) were recognized by their ability to coagulate or flocculate under various treatments such as heat or acid; well-known examples at the start of the nineteenth century included albumen from egg whites, blood serum albumin, fibrin, and wheat gluten. The similarity between the cooking of egg whites and the curdling of milk was recognized even in ancient times; for example, the name *albumen* for the egg-white protein was coined by Pliny the Elder from the Latin *albus ovi* (egg white).

With the advice of Jons Jakob Berzelius, the Dutch chemist Gerhardus Johannes Mulder carried out elemental analyses of common animal and plant proteins. To everyone's surprise, all proteins had nearly the same empirical formula, roughly $C_{400}H_{620}N_{100}O_{120}$ with individual sulfur and phosphorus atoms.

Mulder published his findings in two papers (1837,1838) and hypothesized that there was one basic substance (*Grundstoff*) of proteins, and that it was synthesized by plants and absorbed from them by animals in digestion. Berzelius was an early proponent of this theory and proposed the name "protein" for this substance in a letter dated 10 July 1838

Mulder went on to identify the products of protein degradation such as the amino acid, leucine, for which he found a (nearly correct) molecular weight of 131 Da.

Purifications and Measurements of Mass

The minimum molecular weight suggested by Mulder's analyses was roughly 9 kDa, hundreds of times larger than other molecules being studied. Hence, the chemical structure of proteins (their primary structure) was an active area of research until 1949, when Fred Sanger sequenced insulin. The (correct) theory that proteins were linear polymers of amino acids linked by peptide bonds was proposed independently and simultaneously by Franz Hofmeister and Emil Fischer at the same conference in 1902. However, some scientists were sceptical that such long macromolecules could be stable in solution. Consequently, numerous alternative theories of the protein primary structure were proposed, e.g., the colloidal hypothesis that proteins were assemblies of small molecules, the cyclol hypothesis of Dorothy Wrinch, the diketopiperazine hypothesis of Emil Abderhalden and the pyrrol/piperidine hypothesis of Troensgard (1942). Most of these theories had difficulties in accounting for the fact that the digestion of proteins yielded peptides and amino acids. Proteins were finally shown to be macromolecules of well-defined composition (and not colloidal mixtures) by Theodor Svedberg using analytical ultracentrifugation. The possibility that some proteins are non-covalent associations of such macromolecules was shown by Gilbert Smithson Adair (by measuring the osmotic pressure of hemoglobin) and, later, by Frederic M. Richards in his studies of ribonuclease S. The mass spectrometry of proteins has long been a useful technique for identifying posttranslational modifications and, more recently, for probing protein structure.

Most proteins are difficult to purify in more than milligram quantities, even using the most modern methods. Hence, early studies focused on proteins that could be purified in large quantities, e.g., those of blood, egg white, various toxins, and digestive/metabolic enzymes obtained from slaughterhouses. Many techniques of protein purification were developed during World War II in a project led by Edwin Joseph Cohn to purify blood proteins to help keep soldiers alive. In the late 1950s, the Armour Hot Dog Co. purified 1 kg (= one million milligrams) of pure bovine pancreatic ribonuclease A and made it freely available to scientists around the world. This generous act made RNase A the main protein for basic research for the next few decades, resulting in several Nobel Prizes.

Protein Folding and First Structural Models

The study of protein folding began in 1910 with a famous paper by Henrietta Chick and C. J. Martin, in which they showed that the flocculation of a protein was composed of two distinct processes: the precipitation of a protein from solution was *preceded* by another process called denaturation, in which the protein became much less soluble, lost its enzymatic activity and became more chemically reactive. In the mid-1920s, Tim Anson and Alfred Mirsky proposed that denaturation was a reversible process, a correct hypothesis that was initially lampooned by some scientists as "unboiling the egg". Anson also suggested that denaturation was a two-state ("all-or-none") process, in which one fundamental molecular transition resulted in the drastic changes in solubility, enzymatic activity and chemical reactivity; he further noted that the free energy changes upon denaturation were much smaller than those typically involved in chemical reactions.

In 1929, Hsien Wu hypothesized that denaturation was protein unfolding, a purely conformational change that resulted in the exposure of amino acid side chains to the solvent. According to this (correct) hypothesis, exposure of aliphatic and reactive side chains to solvent rendered the protein less soluble and more reactive, whereas the loss of a specific conformation caused the loss of enzymatic activity. Although considered plausible, Wu's hypothesis was not immediately accepted, since so little was known of protein structure and enzymology and other factors could account for the changes in solubility, enzymatic activity and chemical reactivity. In the early 1960s, Chris Anfinsen showed that the folding of ribonuclease A was fully reversible with no external cofactors needed, verifying the "thermodynamic hypothesis" of protein folding that the folded state represents the global minimum of free energy for the protein.

The hypothesis of protein folding was followed by research into the physical interactions that stabilize folded protein structures. The crucial role of hydrophobic interactions was hypothesized by Dorothy Wrinch and Irving Langmuir, as a mechanism that might stabilize her cyclol structures. Although supported by J. D. Bernal and others, this (correct) hypothesis was rejected along with the cyclol hypothesis, which was disproven in the 1930s by Linus Pauling (among others). Instead, Pauling championed the idea that protein structure was stabilized mainly by hydrogen bonds, an idea advanced initially by William Astbury (1933). Remarkably, Pauling's incorrect theory about H-bonds resulted in his *correct* models for the secondary structure elements of proteins, the alpha helix and the beta sheet. The hydrophobic interaction was restored to its correct prominence by a famous article in 1959 by Walter Kauzmann on denaturation, based partly on work by Kaj Linderstrom-Lang. The ionic nature of proteins was demonstrated by Bjerrum, Weber and Arne

Tiselius, but Linderstrom-Lang showed that the charges were generally accessible to solvent and not bound to each other (1949).

The secondary and low-resolution tertiary structure of globular proteins was investigated initially by hydrodynamic methods, such as analytical ultracentrifugation and flow birefringence. Spectroscopic methods to probe protein structure (such as circular dichroism, fluorescence, near-ultraviolet and infrared absorbance) were developed in the 1950s. The first atomic-resolution structures of proteins were solved by X-ray crystallography in the 1960s and by NMR in the 1980s. As of 2006, the Protein Data Bank has nearly 40,000 atomic-resolution structures of proteins. In more recent times, cryo-electron microscopy of large macromolecular assemblies and computational protein structure prediction of small protein domains are two methods approaching atomic resolution.

RNA

Ribonucleic acid (RNA) is one of the three major macromolecules (along ith DNA and proteins) that are essential for all known forms of life. Like DNA, RNA is made up of a long chain of components called nucleotides. Each nucleotide consists of a nucleobase (sometimes called a nitrogenous base), a ribose sugar, and a phosphategroup. The sequence of nucleotides allows RNA to encode genetic information. For example, someviruses use RNA instead of DNA as their genetic material, and all organisms use messenger RNA(mRNA) to carry the genetic information that directs the synthesis of proteins.

Like proteins, some RNA molecules play an active role in cells by catalyzing biological reactions, controlling gene expression, or sensing and communicating responses to cellular signals. One of these active processes is protein synthesis, a universal function whereby mRNA molecules direct the assembly of proteins on ribosomes. This process uses transfer RNA (tRNA) molecules to deliver amino acids to the ribosome, where ribosomal RNA (rRNA) links amino acids together to form proteins.

The chemical structure of RNA is very similar to that of DNA, with two differences—(a) RNA contains the sugar ribose while DNA contains the slightly different sugar deoxyribose (a type of ribose that lacks one oxygen atom), and (b) RNA has the nucleobase uracil while DNA contains thymine (uracil and thymine have similar base-pairing properties).

Unlike DNA, most RNA molecules are single-stranded. Single-stranded RNA molecules adopt very complex three-dimensional structures, since they are not restricted to the repetitive double-helical form of double-stranded DNA. RNA is made within living cells by RNA polymerases, enzymes which act to copy a DNA or RNA template into a new RNA strand through processes known as transcriptionor RNA replication, respectively.

RNA and DNA are both nucleic acids, but differ in three main ways. First, unlike DNA which is generally double-stranded, RNA is a single-stranded molecule in many of its biological roles and has a much shorter chain of nucleotides. Second, while DNA contains *deoxyribose*, RNA contains *ribose* (there is no hydroxyl group attached to the pentose ring in the 2' position in DNA). These hydroxyl groups make RNA less stable than DNA because it is more prone to hydrolysis. Third, the complementary base to adenine is not thymine, as it is in DNA, but rather uracil, which is anunmethylated form of thymine.

Like DNA, most biologically active RNAs, including mRNA, tRNA, rRNA, snRNAs and other non-coding RNAs, contain self-complementary sequences that allow parts of the RNA to fold and pair with itself to form double helices. Structural analysis of these RNAs has revealed that they are highly structured. Unlike DNA, their structures do not consist of long double helices but rather collections of short helices packed together into structures akin to proteins. In this fashion, RNAs can achieve chemical catalysis, like enzymes. For instance, determination of the structure of the ribosome—an enzyme that catalyses peptide bond formation—revealed that its active site is composed entirely of RNA.

Structure

Each nucleotide in RNA contains a ribose sugar, with carbons numbered 1' through 5'. A base is attached to the 1' position, generally adenine (A), cytosine (C), guanine (G) or uracil (U). Adenine and guanine are purines, cytosine and uracil are pyrimidines. A phosphate group is attached to the 3' position of one ribose and the 5' position of the next. The phosphate groups have a negative charge each at physiological pH, making RNA a charged molecule (polyanion). The bases may form hydrogen bonds between cytosine and guanine, between adenine and uracil and between guanine and uracil. However other interactions are possible, such as a group of adenine bases binding to each other in a bulge, or the GNRA tetraloop that has a guanine–adenine base-pair.n important structural feature of RNA that distinguishes it from DNA is the presence of ahydroxyl group at the 2' position of the ribose sugar. The presence of this functional group causes the helix to adopt the A-form geometry rather than the B-form most commonly observed in DNA. This results in a very deep and narrow major groove and a shallow and wide minor groove. A second consequence of the presence of the 2'-hydroxyl group is that in conformationally flexible regions of an RNA molecule (that is, not involved in formation of a double helix), it can chemically attack the adjacent phosphodiester bond to cleave the backbone. RNA is transcribed with only four bases (adenine, cytosine, guanine and uracil), but these bases and attached sugars can be modified in numerous ways as the RNAs mature. Pseudouridine (y), in which the linkage between uracil and ribose is changed from a C–N bond to a C–C bond, and ribothymidine (T), are found in various places (most notably in the TyC loop of tRNA). Another notable modified base is hypoxanthine, a deaminated adenine base whose nucleoside is called inosine (I). Inosine plays a key role in the wobble hypothesis of thegenetic code.

There are nearly 100 other naturally occurring modified nucleosides, of which pseudouridine and nucleosides with 2'-O-methylribose are the most common. The specific roles of many of these modifications in RNA are not fully understood. However, it is notable that in ribosomal RNA, many of the post-transcriptional modifications occur in highly functional regions, such as the peptidyl transferase centre and the subunit interface, implying that they are important for normal function.

The functional form of single stranded RNA molecules, just like proteins, frequently requires a specific tertiary structure. The scaffold for this structure is provided by secondary structural elements which are hydrogen bonds within the molecule. This leads to several recognizable "domains" of secondary structure like hairpin loops, bulges and internal loops. Since RNA is charged, metal ions such as Mg^{2+} are needed to stabilise many secondary and tertiary structures.

Synthesis of RNA is usually catalysed by an enzyme—RNA polymerase—using DNA as a template, a process known as transcription. Initiation of transcription begins with the binding of the enzyme to a promoter sequence in the DNA (usually found "upstream" of a gene). The DNA double helix is unwound by the helicase activity of the enzyme. The enzyme then progresses along the template strand in the 3' to 5' direction, synthesizing a complementary RNA molecule with elongation occurring in the 5' to 3' direction. The DNA sequence also dictates where termination of RNA synthesis will occur.

RNAs are often modified by enzymes after transcription. For example, a poly(A) tail and a 5' cap are added to eukaryotic pre-mRNA andintrons are removed by the spliceosome.

There are also a number of RNA-dependent RNA polymerases that use RNA as their template for synthesis of a new strand of RNA. For instance, a number of RNA viruses (such as poliovirus) use this type of enzyme to replicate their genetic material. Also, RNA-dependent RNA polymerase is part of the RNA interference pathway in many organisms.

Messenger RNA

Messenger RNA (mRNA) is a molecule of RNA encoding a chemical "blueprint" for a protein product. mRNA is transcribed from a DNA template, and carries coding information to the sites of protein synthesis: the ribosomes. Here, the nucleic acid polymer is translated into a polymer of amino acids: a protein. In mRNA as in DNA, genetic information is encoded in the sequence of nucleotides arranged into codonsconsisting of three bases each. Each codon encodes for a specific amino acid, except the stop codons that terminate protein synthesis. This process requires two other types of RNA: transfer RNA (tRNA) mediates recognition of the codon and provides the corresponding amino acid, while ribosomal RNA (rRNA) is the central component of the ribosome's protein manufacturing machinery.

Synthesis, Processing, and Function

The brief existence of an mRNA molecule begins with transcription and ultimately ends in degradation. During its life, an mRNA molecule may also be processed, edited, and transported prior to translation. Eukaryotic mRNA molecules often require extensive processing and transport, while prokaryotic molecules do not.

Transcription

During transcription, RNA polymerase makes a copy of a gene from the DNA to mRNA as needed. This process is similar in eukaryotes and prokaryotes. One notable difference, however, is that prokaryotic RNA polymerase associates with mRNA processing enzymes during transcription so that processing can proceed quickly after the start of transcription. The short-lived, unprocessed or partially processed, product is termed *pre-mRNA*; once completely processed, it is termed *mature mRNA*.

Eukaryotic Pre-mRNA Processing

Processing of mRNA differs greatly among eukaryotes, bacteria and archea. Non-eukaryotic mRNA is essentially mature upon transcription and requires no processing, except in rare cases. Eukaryotic pre-mRNA, however, requires extensive processing.

5' Cap Addition

A *5' cap* (also termed an RNA cap, an RNA 7-methylguanosine cap or an RNA m^7G cap) is a modified guanine nucleotide that has been added to the "front" or 5' end of a eukaryotic messenger RNA shortly after the start of transcription. The 5' cap consists of a terminal 7-methylguanosine residue which is linked through a 5'-5'-triphosphate bond to the first transcribed nucleotide. Its presence is critical for recognition by the ribosome and protection from RNases.

Cap addition is coupled to transcription, and occurs co-transcriptionally, such that each influences the other. Shortly after the start of transcription, the 5' end of the mRNA being synthesized is bound by a cap-synthesizing complex associated with RNA polymerase. Thisenzymatic complex catalyses the chemical reactions that are required for mRNA capping. Synthesis proceeds as a multi-step biochemicalreaction.

Splicing

Splicing is the process by which pre-mRNA is modified to remove certain stretches of non-coding sequences called introns; the stretches that remain include protein-coding sequences and are called exons. Sometimes pre-mRNA messages may be spliced in several different ways, allowing a single gene to encode multiple proteins. This process is called alternative splicing. Splicing is usually performed by an RNA-protein complex called the spliceosome, but some RNA molecules are also capable of catalyzing their own splicing.

Editing

In some instances, an mRNA will be edited, changing the nucleotide composition of that mRNA. An example in humans is the apolipoprotein B mRNA, which is edited in some tissues, but not others. The editing creates an early stop codon, which upon translation, produces a shorter protein.

Polyadenylation

Polyadenylation is the covalent linkage of a polyadenylyl moiety to a messenger RNA molecule. In eukaryotic organisms, most messenger RNA (mRNA) molecules are polyadenylated at the 3' end. The poly(A) tail and the protein bound to it aid in protecting mRNA from degradation by exonucleases. Polyadenylation is also important for transcription termination, export of the mRNA from the nucleus, and translation. mRNA can also be polyadenylated in prokaryotic organisms, where poly(A) tails act to facilitate, rather than impede, exonucleolytic degradation. Polyadenylation occurs during and immediately after transcription of DNA into RNA. After transcription has been terminated, the mRNA chain is cleaved through the action of an endonuclease complex associated with RNA polymerase. After the mRNA has been cleaved, around 250 adenosine residues are added to the free 3' end at the cleavage site. This reaction is catalysed by polyadenylate polymerase. Just as in alternative splicing, there can be more than one polyadenylation variant of a mRNA.

Transport

Another difference between eukaryotes and prokaryotes is mRNA transport. Because eukaryotic transcription and translation is compartmentally separated, eukaryotic mRNAs must be exported from the nucleus to the cytoplasm. Mature mRNAs are recognized by their processed modifications and then exported through the nuclear pore. In neurons mRNA must be transported from the soma to the dendriteswhere local translation occurs in response to external stimuli. Many messages are marked with so-called "zip codes" which targets their transport to a specific location.

Translation

Because prokaryotic mRNA does not need to be processed or transported, translation by the ribosome can begin immediately after the end of transcription. Therefore, it can be said that prokaryotic translation is *coupled* to transcription and occurs *co-transcriptionally*. Eukaryotic mRNA that has been processed and transported to the cytoplasm (i.e. mature mRNA) can then be translated by the ribosome. Translation may occur at ribosomes free-floating in the cytoplasm, or directed to the endoplasmic reticulum by the signal recognition particle. Therefore, unlike prokaryotes, eukaryotic translation *is not* directly coupled to transcription.

5' cap

The 5' cap is a specially altered nucleotide on the 5' end of precursor messenger RNA and some other primary RNA transcripts as found ineukaryotes and, as a special exception, caliciviruses such as noroviruses. The process of 5' capping is vital to creating mature messenger RNA, which is then able to undergo translation. Capping ensures the messenger RNA's stability while it undergoes translation in the process of protein synthesis, and is a highly regulated process that occurs in the cell nucleus.

5' Cap Structure

The 5' cap is found on the 5' end of an mRNA molecule and consists of a guanine nucleotide connected to the mRNA via an unusual 5' to 5' triphosphate linkage. This guanosine is methylated on the 7 position directly after capping *in vitro* by a methyl transferase. It is referred to as a 7-methylguanosine cap, abbreviated m^7G.

Further modifications include the possible methylation of the 2' hydroxy-groups of the first 3 ribose sugars of the 5' end of the mRNA. The methylation of both 2' hydroxy-groups is shown on the diagram.

The 5' cap looks like the 3' end of an RNA molecule (the 5' carbon of the cap ribose is bonded, and the 3' unbonded). This provides significant resistance to 5' exonucleases.

Capping Process

The starting point is the unaltered 5' end of an RNA molecule. This features a final nucleotide followed by three phosphate groups attached to the 5' carbon.

1. One of the terminal phosphate groups is removed (by a phosphatase), leaving two terminal phosphates.
2. GTP is added to the terminal phosphates (by a guanylyl transferase), losing two phosphate groups (from the GTP) in the process. This results in the 5' to 5' triphosphate linkage.

3. The 7-Nitrogen of guanine is methylated (by a methyl transferase).
4. Other methyltransferases are optionally used to carry out methylation of 5' proximal nucleotides.

5' Capping Targeting

The Capping Enzyme Complex (CEC) required for capping is found bound to the RNA polymerase II before transcription starts. As soon as the 5' end of the new transcript emerges the enzymes transfer to it and begin the capping process (this is a similar kind of mechanism to ensure capping as for polyadenylation).

The enzymes for capping can only bind to RNA polymerase II ensuring specificity to only these transcripts, which are almost entirely mRNA.

5' Cap Function

The 5' cap has 4 main functions:

1. Regulation of nuclear export.
2. Prevention of degradation by exonucleases.
3. Promotion of translation.
4. Promotion of 5' proximal intron excision.

Nuclear export of RNA is regulated by the Cap binding complex (CBC), which binds exclusively to capped RNA. The CBC is then recognized by the nuclear pore complex and exported. Once in the cytoplasm after the pioneer round of translation, the CBC complex is replaced by the translation factors eIF-4E and eIF-4G. This complex is then recognized by other translation initiation machinery including the ribosome. Cap prevents 5' degradation in two ways. First, degradation of the mRNA by 5' exonucleases is prevented (as mentioned above) by functionally looking like a 3' end. Second, the CBC complex and the eIF-4E/eIF-4G block the access of decapping enzymes to the cap. This increases the half-life of the mRNA, essential in eukaryotes as the export process takes significant time.

Decapping of an mRNA is catalysed by the decapping complex made up of at least Dcp1 and Dcp2, which must compete with eIF-4E to bind the cap. Thus the 5' cap is a marker of an actively translating mRNA and is used by cells to regulate mRNA half-lives in response to new stimuli. Undesirable mRNAs are sent to P-bodies for temporary storage or decapping, the details of which are still being resolved.

The mechanism of 5' proximal intron excision promotion is not well understood, but the 5' cap appears to loop around and interact with thespliceosome in the splicing process, promoting intron excision.

Coding Regions

Coding regions are composed of codons, which are decoded and translated (in eukaryotes usually into one and in prokaryotes usually into several) proteins by the ribosome. Coding regions begin with the start codon and end with a stop codon. Generally, the start codon is an AUG triplet and the stop codon is UAA, UAG, or UGA. The coding regions tend to be stabilised by internal base pairs, this impedes degradation. In addition to being protein-coding, portions of coding regions may serve as regulatory sequences in the pre-mRNA asexonic splicing enhancers or exonic splicing silencers.

Five Prime Untranslated Region

The five prime untranslated region (5' UTR), also known as the leader sequence, is a particular section of messenger RNA (mRNA) and the DNA that codes for it. At times the leader sequence can contain elements for control gene expression by way of attenuation.

The leader sequence will be translated and the availability of the encoded amino acids will affect the transcription of the rest of the mRNA. It starts at the +1 position (where transcription begins) and ends one nucleotide before the start codon (usually AUG) of the coding region. It usually contains a ribosome binding site (RBS), in bacteria also known as the Shine Dalgarno sequence (AGGAGGU). The 5' UTR may be a hundred or more nucleotides long, and the 3' UTR may be even longer (up to several kilobases in length)

.

An mRNA molecule codes for a protein through translation. The mRNA also contains regions that are not translated: in eukaryotes this includes the 5' untranslated region, 3' untranslated region, 5' cap and poly-A tail.

In prokaryotic mRNA the 5' UTR is normally short. Some viruses and cellular genes have unusual long structured 5' UTRs which may have roles in gene expression.

Several regulatory sequences may be found in the 5' UTR:

* Binding sites for proteins, that may affect the mRNA's stability or translation, for example iron responsive elements, which occur in the 5' UTRs (and 3' UTRs) of a small number ofeukaryotic mRNAs that regulate gene expression in response to iron.
* Regulatory elements that do not depend on proteins, such asriboswitches.
* Sequences that promote the initiation of translation.

Three Prime Untranslated Region

The three prime untranslated region (3' UTR) is a particular section of messenger RNA (mRNA). It follows the coding region.

An mRNA molecule codes for a protein through translation. The mRNA also contains regions that are not translated. In eukaryotes these regions are the cap, 5' untranslated region, 3' untranslated region, and polyA tail.

In prokaryotes mRNA structures have some differences (seemRNA) as do histone mRNAs. However, both have 3' UTRs.

Several regulatory sequences are found in the 3' UTR:

* A polyadenylation signal, usually *AAUAAA*, or a slight variant. This marks the site of cleavage of the transcript approximately 30 base pairs past the signal, followed by the addition of several hundred adenine residues (poly-A tail).
* Binding sites for proteins, that may affect the mRNA's stability or location in the cell, like SECIS elements (which direct the ribosome to translate the codon UGA as selenocysteines rather than as a stop codon), or AU-rich elements (AREs), stretches consisting of mainlyadenine and uracil nucleotides (which can either stabilize or destabilize the mRNA depending on the protein bound to it).
* Binding sites for miRNAs, a type of RNAi.

Polyadenylation

Polyadenylation is the addition of a poly(A) tail to an RNAmolecule. The poly(A) tail consists of multiple adenosine monophosphates; in other words, it is a stretch of RNA which only has Adenine bases. In eukaryotes, polyadenylation is part of the process that produces mature messenger RNA(mRNA) for translation. It therefore forms part of the larger process of gene expression.

The process of polyadenylation begins as the transcription of a gene finishes. The 3'-most segment of the newly-made RNA is first cleaved off by a set of proteins; these proteins then synthesise the poly(A) tail at the RNA's 3' end. In some genes these proteins may add a poly(A) tail at any one of several possible sites, polyadenylation can therefore produce more than one transcript from a single gene, similar to alternative splicing.

The poly(A) tail is important for the nuclear export, translation and stability of mRNA. The tail is shortened over time and when it is short enough, the mRNA is enzymatically degraded.

However, in a few cell types, mRNAs with short poly(A) tails are stored for later activation by re-polyadenylation in the cytosol. In contrast, when polyadenylation occurs in bacteria, it promotes RNA degradation. This is also sometimes the case for eukaryotic non-coding RNAs. The wide distribution of polyadenylation among living organisms indicates that this process evolved early in the history of life on Earth.

Primer on RNA

RNAs are a type of large biological molecules, whose individual building blocks are called nucleotides. The name *poly(A) tail* (for polyadenylic acid tail) reflects the way RNA nucleotides are abbreviated, with a letter for the base the nucleotide contains (A for adenine, C for cytosine, G for guanine and U for uracil). RNAs are produced (*transcribed*) from a DNA template. By convention, RNA sequences are written in a 5' to 3' direction.

The 5' end is the part of the RNA molecule that is transcribed first, and the 3' end is transcribed last. The 3' end is also where the poly(A) tail is found on polyadenylated RNAs.

Messenger RNA (mRNA) is RNA that has a coding region that acts as a template for protein synthesis (*translation*). The rest of the mRNA, the *untranslated regions*, tune how active the mRNA is. There are also many RNAs that are not translated, called non-coding RNAs. Like the untranslated regions, many of these non-coding RNAs have regulatory roles.

Nuclear Polyadenylation

Function

In nuclear polyadenylation, a poly(A) tail is added to an RNA at the end of transcription. On mRNAs, the poly(A) tail protects the mRNA molecule from enzymatic degradation in the cytoplasm and aids in transcription termination, export of the mRNA from the nucleus, and translation. Almost all eukaryotic mRNAs are polyadenylated, with the exception of animal replication-dependent histone mRNAs. These are the only mRNAs in eukaryotes that lack a poly(A) tail, ending instead in a stem-loop structure followed by a purine-rich sequence, termed histone downstream element, that directs where the RNA is cut so that the 3' end of the histone mRNA is formed.

Many eukaryotic non-coding RNAs are always polyadenylated at the end of transcription. There are small RNAs where the poly(A) tail is only seen in intermediary forms and not in the mature RNA as the ends are removed during processing, notably microRNAs. But for manylong noncoding RNAs – a seemingly large group of regulatory RNAs that for example includes the RNA Xist which mediates X chromosome inactivation – a poly(A) tail is part of the mature RNA.

Mechanism

The polyadenylation machinery in the nucleus of eukaryotes works on products of RNA polymerase II, such as precursor mRNA. Here, a multi-protein complex cleaves the 3'-most part of a newly produced RNA and polyadenylates the end produced by this cleavage. The cleavage is catalysed by the enzyme CPSF and occurs 10–30 nucleotides downstream of its binding site.This site is often the sequence AAUAAA on the RNA, but variants of it exist that bind more weakly to CPSF. Two other proteins add specificity to the binding to an RNA: CstF and CFI. CstF binds to a GU-rich region further downstream of CPSF's site.CFI recognises a third site on the RNA (a set of UGUAA sequences in mammals) and can recruit CPSF even if the AAUAAA sequence is missing. The polyadenylation signal – the sequence motif recognised by the RNA cleavage complex – varies between groups of eukaryotes. Most human polyadenylation sites contain the AAUAAA sequence, but this sequence is less common in plants and fungi.

The RNA is cleaved right after transcription, as CstF also binds to RNA polymerase II. Cleavage also involves the protein CFII, unknown how. The cleavage site associated with a polyadenylation signal can vary up to some 50 nucleotides. When the RNA is cleaved, polyadenylation starts, catalysed by polyadenylate polymerase. Polyadenylate polymerase builds the poly(A) tail by adding adenosine monophosphate units from adenosine triphosphate to the RNA, cleaving off pyrophosphate. Another protein, PAB2, binds to the new, short poly(A) tail and increases the affinity of polyadenylate polymerase for the RNA. When the poly(A) tail is approximately 250 nucleotides long the enzyme can no longer bind to CPSF and polyadenylation stops, thus determing the length of the poly(A) tail. CPSF is in contact with RNA polymerase II, allowing it to tell the polymerase to terminate transcription. The polyadenylation machinery is also physically linked to the spliceosome, a complex that removes introns from RNAs.

Downstream Effects

The poly(A) tail acts as the binding site for poly(A)-binding protein. Poly(A)-binding protein promotes export from the nucleus and translation, and inhibits degradation. This protein binds to the poly(A) tail prior to mRNA export from the nucleus and in yeast also recruits poly(A) nuclease, an enzyme that shortens the poly(A) tail and allows the export of the mRNA. Poly(A)-binding protein is exported to the cytoplasm with the RNA. mRNAs that are not exported are degraded by the exosome. Poly(A)-binding protein also can bind to, and thus recruit, several proteins that affect translation, one of these is initiation factor-4G which in turn recruits the 40Sribosomal subunit. However, a poly(A) tail is not required for the translation of all mRNAs.

Deadenylation

In eukaryotic somatic cells, the poly(A) tail of most mRNAs in the cytoplasm gradually get shorter, and mRNAs with shorter poly(A) tail are translated less and degraded sooner. However, it can take many hours before an mRNA is degraded. This deadenylation and degradation process can be accelerated by microRNAs complementary to the 3' untranslated region of an mRNA. In immature egg cells, mRNAs with shortened poly(A) tails are not degraded, but are instead stored without being translated. They are then activated by cytoplasmic polyadenylation after fertilisation, during egg activation.

In animals, poly(A) ribonuclease (PARN) can bind to the 5' cap and remove nucleotides from the poly(A) tail. The level of access to the 5' cap and poly(A) tail is important in controlling how soon the mRNA is degraded. PARN deadenylates less if the RNA is bound by the initiation factors 4E (at the 5' cap) and 4G (at the poly(A) tail), so this is why translation reduces deadenylation. The rate of deadenylation may also be regulated by RNA-binding proteins. Once the poly(A) tail is removed, the decapping complex removes the 5' cap, leading to a degradation of the RNA. Several other enzymes that seem to be involved in deadenylation have been identified in yeast.

Alternative Polyadenylation

Many protein-coding genes have more than one polyadenylation site, so a gene can code for several mRNAs that differ in their 3' end. Since alternative polyadenylation changes the length of the 3' untranslated region, it can change which binding sites for microRNAs the 3' untranslated region contains. MicroRNAs tend to repress translation and promote degradation of the mRNAs they bind to, although there are examples of microRNAs that stabilise transcripts. Alternative polyadenylation can also shorten the coding region, thus making the mRNA code for a different protein, but this is much less common than just shortening the 3' untranslated region.

The choice of poly(A) site depends on the expression of the proteins that take part in polyadenylation. For example, the expression of CstF-64, a subunit of cleavage stimulatory factor (CstF), increases in macrophages in response to lipopolysaccharides (a group of bacterial compounds that trigger an immune response). This results in the selection of weak poly(A) sites and thus shorter transcripts.

This removes regulatory elements in the 3' untranslated regions of mRNAs for defence-related products like lysozyme and TNF-a. These mRNAs then have longer half-lives and produce more of these proteins. RNA-binding proteins other than those in the polyadenylation machinery can also affect whether a polyadenyation site is used, as can DNA methylationnear the polyadenylation signal.

Cytoplasmic Polyadenylation

There is polyadenylation in the cytosol of some animal cell types, namely in the germ line, during early embryogenesis and in post-synapticsites of nerve cells. This lengthens the poly(A) tail of an mRNA with a shortened poly(A) tail, so that the mRNA will be translated.These shortened poly(A) tails are often less than 20 nucleotides, and are lengthened to around 80–150 nucleotides.

In the early mouse embryo, cytoplasmic polyadenylation of maternal RNAs from the egg cell allows the cell to survive and grow even though transcription does not start until the middle of the 2-cell stage (4-cell stage in human). In the brain, cytoplasmic polyadenylation is active during learning and could to play a role in long-term potentiation, which is the strengthening of the signal transmission from a nerve cell to another in response to nerve impulses and is important for learning and memory formation.

Cytoplasmic polyadenylation requires the RNA-binding proteins CPSF and CPEB, and can involve other RNA-binding proteins like Pumilio. Depending on the cell type, the polymerase can be the same type of polyadenylate polymerase (PAP) that is used in the nuclear process, or the cytoplasmic polymerase GLD-2.

Tagging for Degradation in Eukaryotes

For many non-coding RNAs, including tRNA, rRNA, snRNA and snoRNA, polyadenylation is a way of marking the RNA for degradation, in at least yeast. This polyadenylation is done in the nucleus by the TRAMP complex, which adds a tail that is around 40 nucleotides long to the 3' end. The RNA is then degraded by the exosome. Poly(A) tails have also been found on human rRNA fragments, both the form of homopolymeric (A only) and heterpolymeric (mostly A) tails.

In Prokaryotes and Organelles

In many bacteria, both mRNAs and non-coding RNAs can be polyadenylated. This poly(A) tail promotes degradation by the degradosome, which contains two RNA-degrading enzymes: polynucleotide phosphorylase and RNase E. Polynucleotide phosphorylase binds to the 3' end of RNAs and the 3' extension provided by the poly(A) tail allows it to bind to the RNAs whose secondary structure would otherwise block the 3' end. Successive rounds of polyadenylation and degradation of the 3' end by polynucleotide phosphorylase allows the degradosome to overcome these secondary structures. The poly(A) tail can also recruit RNases that cut the RNA in two. These bacterial poly(A) tails are about 30 nucleotides long.

In as different groups as animals and trypanosomes, the mitochondria contain both stabilising and destabilising poly(A) tails. Destabilising polyadenylation targets both mRNA and noncoding RNAs. The poly(A) tails are 43 nucleotides long on average. The stabilising ones start at the stop codon, and without them the stop codon (UAA) is not complete as the genome only encodes the U or UA part. Plant mitochondria only have destabilising polyadenylation, and yeast mitochondria have no polyadenylation at all.

While many bacteria and mitochondria have polyadenylate polymerases, they also have another type of polyadenylation, performed bypolynucleotide phosphorylase itself. This enzyme is found in bacteria, mitochondria, plastids and as a constituent of the archeal exosome (in those archaea that have an exosome). It can synthesise a 3' extension where the vast majority of the bases are adenines. Like in bacteria, polyadenylation by polynucleotide phosphorylase promotes degradation of the RNA in plastids and likely also archaea.

Evolution

Although polyadenylation is seen in almost all organisms, it is not universal. However, the wide distribution of this modification and the fact that it is present in organisms from all three domains of life implies that the last universal common ancestor of all living organisms probably had some form of polyadenylation system. A few organisms do not polyadenylate mRNA, which implies that they have lost their polyadenylation machineries during evolution. Although no examples of eukaryotes that lack polyadenylation are known, mRNAs from the bacterium *Mycoplasma gallisepticum* and the salt-tolerant archaean *Haloferax volcanii* lack this modification.

The most ancient polyadenylating enzyme is polynucleotide phosphorylase. This enzyme is part of both the bacterial degradosome and archaeal exosome, two closely related complexes that recycle RNA into nucleotides. This enzyme degrades RNA by attacking the bond between the 3'-most nucleotides with a phosphate, breaking off a diphosphate nucleotide. This reaction is reversible, and so the enzyme can also extend RNA with more nucleotides. The heteropolymeric tail added by polynucleotide phosphorylase is very rich in adenine. The choice of adenine is most likely the result of higher ADP concentrations than other nucleotides as a result of using ATP as an energy currency, making it more likely to be incorporated in this tail in early lifeforms. It has been suggested that the involvement of adenine-rich tails in RNA degradation prompted the later evolution of polyadenylate polymerases (the enzymes that produce poly(A) tails with no other nucleotides in them).

Polyadenylate polymerases are not as ancient. They have separately evolved in both bacteria and eukaryotes from CCA-adding enzyme, which is the enzyme that completes the 3' ends of tRNAs. Its catalytic domain is homologous to that of other polymerases. Likely, the horizontal transfer of bacterial CCA-adding enzyme to eukaryotes allowed the archaeal-like CCA-adding enzyme to switch function to a poly(A) polymerase. Some lineages, like archaea and cyanobacteria, never evolved a polyadenylate polymerase.

History

Polyadenylation was first identified in 1960 as an enzymatic activity in extracts made from cell nuclei that could polymerise ATP, but not ADP, into polyadenine. Although identified in many types of cells, this activity had no known function until 1971, when poly(A) sequences were found in mRNAs.

Initially, the only function of these sequences was thought to be protection of the 3' end of the RNA from nucleases, but later the specific roles of polyadenylation in nuclear export and translation were identified. The polymerases responsible for polyadenylation were first purified and characterized in the 1960s and 1970s, but the large number of accessory proteins that control this process were only discovered in the early 1990s.

Monocistronic Versus Polycistronic mRNA

An mRNA molecule is said to be monocistronic when it contains the genetic information to translate only a single protein. This is the case for most of the eukaryotic mRNAs. On the other hand, polycistronic mRNA carries the information of several genes, which are translated into several proteins. These proteins usually have a related function and are grouped and regulated together in an operon. Most of the mRNA found in bacteria and archea are polycistronic. Dicistronic or bicistronic is the term used to describe an mRNA that encodes only twoproteins.

mRNA Circularization

In eukaryotes it is thought that mRNA molecules form circular structures due to an interaction between the cap binding complex and poly(A)-binding protein. Circularization is thought to promote recycling of ribosomes on the same message leading to efficient translation.

Degradation

Different mRNAs within the same cell have distinct lifetimes (stabilities). In bacterial cells, individual mRNAs can survive from seconds to more than an hour; in mammalian cells, mRNA lifetimes range from several minutes to days. The greater the stability of an mRNA, the more protein may be produced from that mRNA. The limited lifetime of mRNA enables a cell to alter protein synthesis rapidly in response to its changing needs. There are many mechanisms that lead to the destruction of a mRNA, some of which are described below.

Prokaryotic mRNA Degradation

In prokaryotes the lifetime of mRNA is generally much shorter than in eukaryotes. Prokaryotes degrade messages by using a combination of ribonucleases, including endonucleases, 3' exonucleases, and 5' exonucleases. In some instances, small RNA molecules (sRNA) tens to hundreds of nucleotides long can stimulate the degradation of specific mRNAs by base pairing with complementary sequences and facilitating ribonuclease cleavage. It was recently shown that bacteria also have a sort of 5' cap consisting of a triphosphate on the 5' end.Removal of two of the phosphates leaves a 5' monophosphate, causing the message to be destroyed by the endonuclease RNase E.

Eukaryotic mRNA Turnover

Inside eukaryotic cells there is a balance between the processes of translation and mRNA decay. Messages that are being actively translated are bound by ribosomes, the eukaryotic initiation factors eIF-4E and eIF-4G, and poly(A)-binding protein. eIF-4E and eIF-4G block the decapping enzyme (DCP2), and poly(A)-binding protein blocks the exosome complex, protecting the ends of the message. The balance between translation and decay is reflected in the size and abundance of cytoplasmic structures known as P-bodies.

The poly(A) tail of the mRNA is shortened by specialized exonucleases that are targeted to specific messenger RNAs by a combination of cis-regulatory sequences on the RNA and trans-acting RNA-binding proteins. Poly(A) tail removal is thought to disrupt the circular structure of the message and destabilize the cap binding complex. The message is then subject to degradation by either the exosome complex or the decapping complex. In this way, translationally inactive messages can be destroyed quickly, while active messages remain intact. The mechanism by which translation stops and the message is handed-off to decay complexes is not understood in detail.

AU-rich Element Decay

The presence of AU-rich elements in some mammalian mRNAs tends to destabilize those transcripts through the action of cellular proteins that bind these sequences and stimulate poly(A) tail removal. Loss of the poly(A) tail is thought to promote mRNA degradation by facilitating attack by both the exosome complex and the decapping complex. Rapid mRNA degradation via AU-rich elements is a critical mechanism for preventing the overproduction of potent cytokines such as tumour necrosis factor (TNF) and granulocyte-macrophage colony stimulating factor (GM-CSF). AU-rich elements also regulate the biosynthesis of proto-oncogenic transcription factors like c-Jun and c-Fos.

Nonsense Mediated Decay

Nonsense mediated decay (NMD) is a cellular mechanism of mRNA surveillance that functions to detect nonsense mutations and prevent the expression of truncated or erroneous proteins. Following transcription, precursor mRNA undergoes an assemblage of ribonucleoprotein(RNP) components followed by regulatory pre-mRNA processing.

Large average intron size in eukaryotic cells greatly increases the probability that aberrant mRNA splicing will result in the presence of a nonsense (stop) codon (UAA, UAG, UGA) somewhere within the open reading frame. NMD is triggered by exon junction complexes (EJCs) (components of the assembled RNP) that are deposited during pre-mRNA processing. While EJCs located in an open reading frame upstream of exon-exon junctions may facilitate ribosomal recruitmentprior to displacement by a "pioneer" round of translation, EJCs located downstream of a nonsense codon are not displaced because the ribosome is released from the transcript before reaching it. These remaining EJCs function as tags for recruitment of UPF1 following the mRNA's transport out of the nucleus and into the cytosol where the RNA is degraded, for example by the exosome complex. NMD is not only a mechanism for degrading aberrant mRNA, however, as there are numerous examples of normal transcripts whose expression is regulated by this process including the plasticity protein Arc/Arg3.1.

Small Interfering RNA

Small interfering RNA (siRNA), sometimes known as short interfering RNA orsilencing RNA, is a class of double-stranded RNA molecules, 20-25 nucleotides in length, that play a variety of roles in biology. Most notably, siRNA is involved in theRNA interference (RNAi) pathway, where it interferes with the expression of a specific gene. In addition to their role in the RNAi pathway, siRNAs also act in RNAi-related pathways, e.g., as an antiviral mechanism or in shaping the chromatinstructure of a genome; the complexity of these pathways is only now being elucidated.

siRNAs were first discovered by David Baulcombe's group at the Sainsbury Laboratory in Norwich, England, as part of post-transcriptional gene silencing(PTGS) in plants. The group published their findings in *Science* in a paper titled "A species of small antisense RNA in posttranscriptional gene silencing in plants".Shortly thereafter, in 2001, synthetic siRNAs were shown to be able to induce RNAi in mammalian cells by Thomas Tuschl, and colleagues in a paper published in*Nature*. This discovery led to a surge in interest in harnessing RNAi for biomedical research and drug development.

Structure

siRNAs have a well-defined structure: a short (usually 21-nt) double strand RNA (dsRNA) with 2-nt 3' overhangs on either end:

Each strand has a 5' phosphate group and a 3' hydroxyl (-OH) group. This structure is the result of processing by dicer, an enzyme that converts either long dsRNAs or small hairpin RNAs into siRNAs. siRNAs can also be exogenously (artificially) introduced into cells by various transfection methods to bring about the specific knockdown of a gene of interest. Essentially any gene of which the sequence is known can thus be targeted based on sequence complementarity with an appropriately tailored siRNA. This has made siRNAs an important tool for gene function and drug target validation studies in the post-genomic era.

RNAi Induction using SiRNAs or their Biosynthetic Precursors

Transfection of an exogenous siRNA can be problematic because the gene knockdown effect is only transient, particularly in rapidly dividing cells. One way of overcoming this challenge is to modify the siRNA in such a way as to allow it to be expressed by an appropriate vector, e.g., aplasmid. This is done by the introduction of a loop between the two strands, thus producing a single transcript, which can be processed into a functional siRNA. Such transcription cassettes typically use an RNA polymerase III promoter (e.g., U6 or H1), which usually directs the transcription of small nuclear RNAs (snRNAs) (U6 is involved in gene splicing; H1 is the RNasecomponent of human RNase P). It is assumed (although not known for certain) that the resulting siRNA transcript is then processed by Dicer.

Innate Immunity

Introduction of too much siRNA can result in nonspecific events due to activation of innate immune responses. Most evidence to date suggests that this is probably due to activation of the dsRNA sensor PKR, although retinoic acid inducible gene I (RIG-I) may also be involved. The induction of cytokines via toll-like receptor 7 (TLR7) has also been described. One promising method of reducing the nonspecific effects is to convert the siRNA into a microRNA. MicroRNAs occur naturally, and by harnessing this endogenous pathway it should be possible to achieve similar gene knockdown at comparatively low concentrations of resulting siRNAs. This should minimize nonspecific effects.

Off-targeting

Off-targeting is another challenge to the use of siRNAs as a gene knockdown tool. Here, genes with incomplete complementarity are inadvertently downregulated by the siRNA (effectively, the siRNA acts as a miRNA), leading to problems in data interpretation and potential toxicity. This, however, can be partly addressed by designing appropriate control experiments, and siRNA design algorithms are currently being developed to produce siRNAs free from off-targeting. Genome-wide expression analysis, e.g., by microarray technology, can then be used to verify this and further refine the algorithms. A 2006 paper from the laboratory of Dr. Khvorova implicates 6- or 7-basepair-long stretches from position 2 onward in the siRNA matching with 3'UTR regions in off-targeted genes.

Possible Therapeutic Applications and Challenges

Given the ability to knock down essentially any gene of interest, RNAi via siRNAs has generated a great deal of interest in both basic and applied biology. There are an increasing number of large-scale RNAi screens that are designed to identify the important genes in various biological pathways. Because disease processes also depend on the activity of multiple genes, it is expected that in some situations turning off the activity of a gene with an siRNA could produce a therapeutic benefit.

However, applying RNAi via siRNAs to living animals, especially humans, poses many challenges. Experimentally, siRNAs show different effectiveness in different cell types in a manner as yet poorly understood: some cells respond well to siRNAs and show a robust knockdown, whereas others show no such knockdown (even despite efficient transfection). Phase I results of the first two therapeutic RNAi trials (indicated for age-related macular degeneration, aka AMD) reported at the end of 2005 that siRNAs are well tolerated and have suitable pharmacokinetic properties.

MicroRNA

MicroRNAs (miRNAs) are short ribonucleic acid (RNA) molecules, on average only 22 nucleotides long and are found in all eukaryotic cells. miRNAs are post-transcriptional regulators that bind to complementary sequences on target messenger RNA transcripts (mRNAs), usually resulting in translational repression and gene silencing. The human genome may encode over 1000 miRNAs, which may target about 60% of mammalian genes and are abundant in many human cell types.

miRNAs show very different characteristics between plants and metazoans. In plants the miRNA complementarity to its mRNA target is nearly perfect, with no or few mismatched bases. In metazoans on the other hand miRNA complementarity is far from perfect and one miRNA can target many different sites on the same mRNA or on many different mRNAs.

Another difference is the location of target sites on mRNAs. In metazoans the miRNA target sites are in the three prime untranslated regions (3'UTR) of the mRNA. In plants targets can be located in the 3' UTR but are more often in the coding region itself.MiRNAs are well conserved in eukaryotic organism and are thought to be a vital and evolutionary ancient component of genetic regulation.

The first miRNAs were characterized in the early 1990s, but miRNAs were not recognized as a distinct class of biologic regulators with conserved functions until the early 2000s. Since then, miRNA research has revealed multiple roles in negative regulation (transcript degradation and sequestering, translational suppression) and possible involvement in positive regulation (transcriptional and translational activation). By affecting gene regulation, miRNAs are likely to be involved in most biologic processes. Different sets of expressed miRNAs are found in different cell types and tissues. Aberrant expression of miRNAs has been implicated in numerous disease states, and miRNA-based therapies are under investigation.

MicroRNAs were discovered in 1993 by Victor Ambros, Rosalind Lee and Rhonda Feinbaum during a study of the gene *lin-14* in *C. elegans*development.

They found that LIN-14 protein abundance was regulated by a short RNA product encoded by the *lin-4* gene. A 61 nucleotide precursor from *lin-4* gene matured to a 22 nucleotide RNA containing sequences partially complementary to multiple sequences in the 3' UTR of the *lin-14* mRNA. This complementarity was sufficient and necessary to inhibit the translation of *lin-14* mRNA into LIN-14 protein. Retrospectively, the *lin-4* small RNA was the first microRNA to be identified, though at the time, it was thought to be a nematode idiosyncrasy. Only in 2000 was a second RNA characterized: let-7, which repressed *lin-41*, *lin-14*, *lin-28*, *lin-42*, and *daf-12* expression during developmental stage transitions in *C. elegans*. let-7 was soon found to be conserved in many species, indicating the existence of a wider phenomenon.

Nomenclature

Under a standard nomenclature system, names are assigned to experimentally confirmed miRNAs before publication of their discovery.The prefix "mir" is followed by a dash and a number, the latter often indicating order of naming. For example, mir-123 was named and likely discovered prior to mir-456.

The uncapitalized "mir-" refers to the pre-miRNA, while a capitalized "miR-" refers to the mature form. miRNAs with nearly identical sequences bar one or two nucleotides are annotated with an additional lower case letter.

For example, miR-123a would be closely related to miR-123b. Pre-miRNAs that lead to 100% identical mature miRNAs but that are located at different places in the genome are indicated with an additional dash-number suffix. For example, the pre-miRNAs hsa-mir-194-1 and hsa-mir-194-2 lead to an identical mature miRNA (hsa-miR-194) but are located in different regions of the genome. Species of origin is designated with a three-letter prefix, e.g., hsa-miR-123 would be from human (*Homo sapiens*) and oar-miR-123 would be a sheep (*Ovis aries*) miRNA.

Other common prefixes include 'v' for viral (miRNA encoded by a viral genome) and 'd' for *Drosophila* miRNA (a fruit fly commonly studied in genetic research). When two mature microRNAs originate from opposite arms of the same pre-miRNA, they are denoted with a -3p or -5p suffix. (In the past, this distinction was also made with 's' (sense) and 'as' (antisense)). When relative expression levels are known, an asterisk following the name indicates an miRNA expressed at low levels relative to the miRNA in the opposite arm of a hairpin. For example, miR-123 and miR-123* would share a pre-miRNA hairpin, but more miR-123 would be found in the cell.

Biogenesis

Most microRNA genes are found in intergenic regions or in anti-sense orientation to genes and contain their own miRNA gene promoter and regulatory units. As much as 40% of miRNA genes may lie in the introns of protein and non-protein coding genes or even in exons. These are usually, though not exclusively, found in a sense orientation and thus usually are regulated together with their host genes.

Other miRNA genes showing a common promoter include the 42-48% of all miRNAs originating from polycistronic units containing 2-7 discrete loops from which mature miRNAs are processed, although this does not necessarily mean the mature miRNAs of a family will be homologous in structure and function. The promoters mentioned have been shown to have some similarities in their motifs to promoters of other genes transcribed by RNA polymerase II such as protein coding genes. The DNA template is not the final word on mature miRNA production: 6% of human miRNAs show RNA editing, the site-specific modification of RNA sequences to yield products different from those encoded by their DNA. This increases the diversity and scope of miRNA action beyond that implicated from the genome alone.

Transcription

miRNA genes are usually transcribed by RNA polymerase II (Pol II). The polymerase often binds to a promoter found near the DNA sequence encoding what will become the hairpin loop of the pre-miRNA. The resulting transcript is capped with a specially-modified nucleotide at the 5' end, polyadenylated with multiple adenosines (a poly(A) tail), and spliced.

The product, called a primary miRNA (pri-miRNA), may be hundreds or thousands of nucleotides in length and contain one or more miRNA stem loops. When a stem loop precursor is found in the 3' UTR, a transcript may serve as a pri-miRNA and a mRNA. RNA polymerase III (Pol III) transcribes some miRNAs, especially those with upstream Alu sequences, transfer RNAs (tRNAs), and mammalian wide interspersed repeat (MWIR) promoter units.

Nuclear Processing

A single pri-miRNA may contain from one to six miRNA precursors. These hairpin loop structures are composed of about 70 nucleotides each. Each hairpin is flanked by sequences necessary for efficient processing.

The double-stranded RNA structure of the hairpins in a pri-miRNA is recognized by a nuclear protein known as DiGeorge Syndrome Critical Region 8 (DGCR8 or "Pasha" in invertebrates), named for its association with DiGeorge Syndrome. DGCR8 associates with the enzyme Drosha, a protein that cuts RNA, to form the "Microprocessor" complex.

In this complex, DGCR8 orients the catalytic RNase III domain of Drosha to liberate hairpins from pri-miRNAs by cleaving RNA about eleven nucleotides from the hairpin base (two helical RNA turns into the stem). The resulting hairpin, known as a pre-miRNA (precursor-miRNA), has a two-nucleotide overhang at its 3' end; it has 3' hydroxyl and 5' phosphate groups.

Pre-miRNAs that are spliced directly out of introns, bypassing the Microprocessor complex, are known as "mirtrons." Originally thought to exist only in *Drosophila* and *C. elegans*, mirtrons have now been found in mammals.

Perhaps as many as 16% of pri-miRNAs may be altered through nuclear RNA editing. Most commonly, enzymes known as adenosine deaminases acting on RNA (ADARs) catalyse adenosine to inosine (A to I) transitions. RNA editing can halt nuclear processing (for example, of pri-miR-142, leading to degradation by the ribonuclease Tudor-SN) and alter downstream processes including cytoplasmic miRNA processing and target specificity (e.g., by changing the seed region of miR-376 in the central nervous system).

Nuclear Export

pre-miRNA hairpins are exported from the nucleus in a process involving the nucleocytoplasmic shuttle Exportin-5. This protein, a member of the *karyopherin* family, recognizes a two-nucleotide overhang left by the RNase III enzyme Drosha at the 3' end of the pre-miRNA hairpin. Exportin-5-mediated transport to the cytoplasm is energy-dependent, using GTP bound to the Ran protein.

Cytoplasmic Processing

In the cytoplasm, the pre-miRNA hairpin is cleaved by the RNase III enzyme Dicer. This endoribonuclease interacts with the 3' end of the hairpin and cuts away the loop joining the 3' and 5' arms, yielding an imperfect miRNA:miRNA* duplex about 22 nucleotides in length.

Overall hairpin length and loop size influence the efficiency of Dicer processing, and the imperfect nature of the miRNA:miRNA* pairing also affects cleavage. Although either strand of the duplex may potentially act as a functional miRNA, only one strand is usually incorporated into the RNA-induced silencing complex (RISC) where the miRNA and its mRNA target interact.

Biogenesis in Plants

miRNA biogenesis in plants differs from metazoan biogenesis mainly in the steps of nuclear processing and export. Instead of being cleaved by two different enzymes, once inside and once outside the nucleus, both cleavages of the plant miRNA is performed by a Dicer homolog, called Dicer-like1 (DL1).

DL1 is only expressed in the nucleus of plant cells, which indicates that both reactions take place inside the nucleus. Before plant miRNA:miRNA* duplexes are transported out of the nucleus its 3' overhangs are methylated by a RNA methyltransferaseprotein called Hua-Enhancer1 (HEN1). The duplex is then transported out of the nucleus to the cytoplasm by a protein called Hasty (HST), an Exportin 5 homolog, where they disassemble and the mature miRNA is incorporated into the RISC.

RNA-induced Silencing Complex

RNA-Induced Silencing Complex, or RISC, is a multiprotein complex that incorporates one strand of a small interfering RNA (siRNA) ormicro RNA (miRNA). RISC uses the siRNA or miRNA as a template for recognizing complementary mRNA. When it finds a complementary strand, it activates RNase and cleaves the RNA. This process is important both in gene regulation by microRNAs and in defence against viral infections, which often use double-stranded RNA as an infectious vector.

Binding

The RISC complex with a bound siRNA recognizes complementary messenger RNA (mRNA) molecules and degrades them, resulting in substantially decreased levels of protein translation and effectively turning off the gene. It is as yet unclear how the activated RISC complex locates the mRNA targets in the cell, though it has been shown that the process can occur in situations outside of ongoing protein translation from mRNA. Endogenously expressed miRNA is usually imperfectly complementary to a large number of nuclear genes and has a modulating effect on these genes' levels of expression via translational repression.

Bibliography

Annie Tan: *Blueprints Notes and Cases - Biochemistry, Genetics and Embryology*, Blackwell Science, Delhi, 2004.

Balkwill, Fran: *DNA Is Here to Stay*, Carolrhoda Books, Inc., Minneapolis, 1993.

Banerjee, S : *Cell Biochemistry*, Dominant Pub, Delhi, 2009.

Bhatia, A L : *Biochemistry and Endocrinology*, Indus Valley, Delhi, 2002.

Brookes, Martin: *Get a Grip on Genetics*, *Time Life Books*, East Sussex, England, 1998.

Bruce, Donald : *Engineering Genesis: The Ethics of Genetic Engineering in Non-Human Animals*, London: Earthscan, 1998.

Cahill, Lisa: *Genetics, Theology, and Ethics: An Interdisciplinary Conversation*, New York: Crossroad, 2005.

Curtis, M.: *The Geometry of DNA: A Structural Revision*. London: Blue Gallery, 1996.

Diaz, E.: *Microbial Biodegradation: Genomics and Molecular Biology,* Caister Academic Press, UK, 2008.

Drlica, Karl : *Understanding DNA and Gene Cloning: A Guide for the Curious*, New York: Wiley, 1997.

Fukuyama F.: *Our Posthuman Future: Consequences of the Biotechnology Revolution,* Farrar, Straus and Giroux, NewYork, 2002.

Garza-Valdes, L.A.: *The DNA of God?* Hodder and Stoughton, London, 1998.

Hames, B. D.: *Genes and Embryos*, Oxford University Press, New York, 1989.

Jonoska, N.: *Self-Assembling DNA Graphs, DNA-Based Computers VIII*, Springer-Verlag, Berlin, 2003.

Kallenbach, N.R.: *Molecular Structure: Chemical Reactivity and Biological Activity,* Oxford University Press, Oxford, 1988.

Lewis, Ricki: *Human Genetics: Concepts and Applications,* Boston, Mass: McGraw Hill. 1994.

Mayr, E.: *Change of Genetic Environment and Evolution*, Allen and Unwin, London, 1954.

Nisha Khalsa: *Essentials of Biochemistry*, Aavishkar Pub, Delhi, 2008.

Old, R.W.: *Principles of Gene Manipulation*, London, Blackwell Scientific Publications, 1989.

Rabus, R: *Microbial Biodegradation: Genomics and Molecular Biology*, Caister Academic Press, UK, 2008.

Sturtevant, Alfred: *History of Genetics*, Harper and Row, New York, 1965.

Stwertka, Eve : *Genetic Engineering*, Franklin Watts, New York, 1982.

Stwertka, Eve and Albert: *Genetic Engineering*, Franklin Watts, New York, 1982.

Vasudev Bhatnagar: *Cell Science and Technology*, Campus Books, Delhi, 2009.

William, Kovacs D.: *An Introduction to Geotechnical Engineering*, Prentice Hall, UK, 1981.

Williams, J.G. and R.K. Patient : *Genetic Engineering*. Oxford, England: IRL Press, 1988.

Wilson, I.: *The Blood and The Shroud*. London: Weidenfeld & Nicolson, 1998.

Winston, R.: *The Future of Genetic Manipulation*, Phoenix, London, 1997.

Index

❑❑❑